# STMODERN GEOGRAPHIES

## THE REASSERTION OF SPACE IN CRITICAL SOCIAL THEORY

## Edward W. Soja

# Postmodern Geographies

## The Reassertion of Space in Critical Social Theory

EDWARD W. SOJA

**V**

**VERSO**

London · New York

This edition published by Verso 1989
© 1989 Edward W. Soja

**Verso**
UK: 6 Meard Street, London W1V 3HR
USA: 29 West 35th Street, New York, NY 10001-2291

Verso is the imprint of New Left Books

**British Library Cataloguing in Publication Data**
Soja, Edward W.
    Postmodern geographies : the reassertion
    of space in critical social theory.
    (Haymarket).
    1. Human geography. Spatial aspects
    I. Title          II. Series
    304.2

    ISBN 0-86091-225-6
    ISBN 0-86091-936-6 Pbk

**US Library of Congress Cataloging in Publication Data**
Soja, Edward W.
    Postmodern geographies : the reassertion of space in critical
    social theory / Edward W. Soja.
        P.    cm. — (Haymarket series)
    Bibliography: p.
    Includes index.
        ISBN 0-86091-225-6 (U.S.) : $50.00.    ISBN 0-86091-936-6 (U.S. :
    pbk.) : $16.95
        1. Geography—Philosophy.       I. Title.
    G70.S62 1988
    910'.01—dc19

Typeset by Leaper & Gard Ltd, Bristol, England
Printed in Great Britain by Bookcraft (Bath) Ltd

# Contents

# Preface and Postscript

Combining a Preface with a Postscript seems a particularly apposite way to introduce (and conclude) a collection of essays on postmodern geographies. It signals right from the start an intention to tamper with the familiar modalities of time, to shake up the normal flow of the linear text to allow other, more 'lateral' connections to be made. The discipline imprinted in a sequentially unfolding narrative predisposes the reader to think historically, making it difficult to see the text as a map, a geography of simultaneous relations and meanings that are tied together by a spatial rather than a temporal logic. My aim is to spatialize the historical narrative, to attach to *durée* an enduring critical human geography.

Each essay in this volume is a different evocation of the same central theme: the reassertion of a critical spatial perspective in contemporary social theory and analysis. For at least the past century, time and history have occupied a privileged position in the practical and theoretical consciousness of Western Marxism and critical social science. Understanding how history is made has been the primary source of emancipatory insight and practical political consciousness, the great variable container for a critical interpretation of social life and practice. Today, however, it may be space more than time that hides consequences from us, the 'making of geography' more than the 'making of history' that provides the most revealing tactical and theoretical world. This is the insistent premise and promise of postmodern geographies.

The essays presented here can, of course, be read in sequence as a textual unfolding of an essentially historical argument. But at the very core of each essay is an attempt to deconstruct and recompose the rigidly historical narrative, to break out from the temporal prisonhouse of language and the similarly carceral historicism of conventional critical theory to make room for the insights of an interpretive human

*1*

geography, a spatial hermeneutic. The sequential flow is thus frequently side-tracked to take coincident account of simultaneities, lateral mappings that make it possible to enter the narration at almost any point without losing track of the general objective: to create more critically revealing ways of looking at the combination of time and space, history and geography, period and region, sequence and simultaneity. Bracketing a preface with a postscript, forewords that are also afterwords, is only the first playful signal of this intentional rebalancing act.

Having started with a twisting of the temporal order, it seems appropriate to suggest that the best introduction to postmodern geographies is represented in the last chapter, a free-wheeling essay on Los Angeles that both integrates and disintegrates what has gone before it. 'Taking Los Angeles Apart' is an inquisitive reading of a decidedly postmodern landscape, a search for revealing 'other spaces' and hidden geographical texts. The essay feeds from Jorge Luis Borges's brilliant sighting/siting of 'The Aleph' – the only place on earth where all places are, a limitless space of simultaneity and paradox, impossible to describe in less than extra-ordinary language. Borges's observations crystallize some of the dilemmas confronting the interpretation of postmodern geographies:

> Then I saw the Aleph ... And here begins my despair as a writer. All language is a set of symbols whose use among its speakers assumes a shared past. How, then, can I translate into words the limitless Aleph, which my floundering mind can scarcely encompass? ... Really, what I want to do is impossible, for any listing of an endless series is doomed to be infinitesimal. In that single gigantic instant I saw millions of acts both delightful and awful; not one of them amazed me more than the fact that all of them occupied the same point in space, without overlapping or transparency. What my eyes beheld was simultaneous, but what I shall now write down will be successive, because language is successive. Nonetheless, I will try to recollect what I can.

Every ambitious exercise in critical geographical description, in translating into words the encompassing and politicized spatiality of social life, provokes a similar linguistic despair. What one sees when one looks at geographies is stubbornly simultaneous, but language dictates a sequential succession, a linear flow of sentential statements bound by that most spatial of earthly constraints, the impossibility of two objects (or words) occupying the same precise place (as on a page). All that we can do is re-collect and creatively juxtapose, experimenting with assertions and insertions of the spatial against the prevailing grain of time. In the end, the interpretation of postmodern geographies can be no more than a beginning.

Backing up this experimental essay is a condensed chapter which maps the political economy of urban restructuring as screened through the postfordist landscapes of contemporary Los Angeles. A more concrete regional geography is presented to exemplify the rise of a new regime of 'flexible' capitalist accumulation tautly built upon a restorative spatial 'fix' and edgily connected to the postmodern cultural fabric. This symptomatic description is followed/preceded by a deeper framing of the historical geography of capitalism via analyses of the evolution of urban form in the capitalist city, the changing mosaics of uneven regional development within the capitalist state, and the various reconfigurations of an international spatial division of labour.

Here, as elsewhere in the text, there is an underlying assumption about the spatio-temporal rhythm of capitalist development, a macrospective conjunction of periodicity and spatialization that is induced from the successful survival of capitalist societies over the past two hundred years. Again the aim is to open up and explore a critical viewpoint that pointedly flows from the resonant interplay of temporal succession and spatial simultaneity. Postmodern and postfordist geographies are defined as the most recent products of a sequence of spatialities that can be complexly correlated to successive eras of capitalist development. I adapt the theory of 'long waves' from the work of Ernest Mandel, Eric Hobsbawm, David Gordon, and others as a revealing spatio-temporal subtext from which to interpret the historical geography of cities, regions, states, and the world economy.

The more empirically-based spatializations of the last three essays are reproduced and explained in a different way in the first two chapters, which locate other postmodern geographies in the wake of a deep restructuring of modern critical social theory and discourse. Appropriating insights from Michel Foucault, John Berger, Fredric Jameson, Ernest Mandel and Henri Lefebvre, I attempt to spatialize the conventional narrative by recomposing the intellectual history of critical social theory around the evolving dialectics of space, time, and social being: geography, history, and society. In the first chapter, the subordination of a spatial hermeneutic is traced back to the nineteenth century origins of historicism and the consequent development of Western Marxism and critical social science, a history periodized by dramatic changes in the conceptualization and experience of modernity. The same crisis-induced rhythm that ripples through the macro-historical geography of capitalist cities and regions is seen reflected in the history of critical theoretical consciousness, creating an interlocking sequence of 'regimes' of critical thought that follows in roughly the same half-century blocks that have phased the changing political economy of capitalism since the age of revolution, the first of four punctuating periods of restructuring and modernization.

The middle half of the nineteenth century, hingeing around the far-reaching events of 1848–51, was the classical era of competitive industrial capitalism. It was also a period in which historicity and spatiality were in approximate balance as sources of emancipatory consciousness, whether one traced critical thought through the perspectives of French socialism, English political economy, or German idealist philosophy. Challenging the specific geography of industrial capitalism, its spatial and territorial structures, was a vital part of the radical critiques and regional social movements arising during this period, just as reforming this geography became an important instrumental objective for the newly entrenched bourgeois states of Europe and North America. After the fall of the Paris Commune, however, the explicitly spatial critiques, radical and liberal, began to recede behind more powerful Eurocentric assertions of the revolutionary subjectivity of time and history.

The last decades of the nineteenth century, examined in retrospect, can be seen as an era of rising historicism and the parallel submergence of space in critical social thought. The socialist critique consolidated around the historical materialism of Marx while a mix of Comtean and neo-Kantian influences reshaped liberal social philosophy and provoked the formation of new 'social sciences' equally determined to understand the development of capitalism as an historical, but only incidentally geographical, process. This rise of a despatializing historicism, only now beginning to be recognized and examined, coincided with the second modernization of capitalism and the onset of an age of empire and corporate oligopoly. It so successfully occluded, devalued, and de-politicized space as an object of critical social discourse that even the possibility of an emancipatory spatial praxis disappeared from view for almost a century.

Little changed with regard to the theoretical primacy of history over geography during the third modernization of capitalism and the ensuing era of Fordism and bureaucratic state-management, stretching roughly from the Russian Revolution to the late 1960s. The nineteenth-century obsession with time and history, as Foucault would call it, continued to bracket modern critical thought. The first chapter begins and ends with Foucault's summative observation: 'Space was treated as the dead, the fixed, the undialectical, the immobile. Time, on the contrary, was rich-ness, fecundity, life, dialectic.' Little eddies of a lively geographical imagination survived outside the mainstreams of Marxism-Leninism and positivist social science, but they were difficult to comprehend and remained decidedly peripheral.

In the late 1960s, however, with the onset of a crisis-induced fourth modernization, this long-lasting modern critical tradition began to change. Both Western Marxism and critical social science appeared to explode into more heterogeneous fragments, losing much of their separate cohesivenesses and centralities. And as we approach another *fin de siècle*, alternative modern movements have appeared to compete for control over the perils and possibilities emerging in a restructured contemporary world. Although they remain controversial and confusing terms, filled with disparate and often disparaging connotations, postmodernity, postmodernization, and postmodernism now seem to be appropriate ways of describing this contemporary cultural, political, and theoretical restructuring; and of highlighting the reassertion of space that is complexly intertwined with it.

Initially suspicious of too hasty a 'rush to the post', I once toyed with the idea of creating a new journal called *Antipost* to do battle not only with postmodernism but also with the multiplying array of other post-prefixed 'isms', from postindustrialism to poststructuralism. I am now, as is obvious from my titular commitment, more comfortable with the epithetic label postmodern and its intentional announcement of a possibly epochal transition in both critical thought and material life. I continue to see the present period primarily as another deep and broad restructuring of modernity rather than as a complete break and replacement of all progressive, post-Enlightenment thought, as some who call themselves postmodernists (but are probably better described as anti-modernists) proclaim. I also understand the suspicious antagonism of the modern left to the presently dominant neo-conservatism and obfuscating whimsy of most postmodern movements. But I am convinced that too many opportunities are missed by dismissing postmodernism as irretrievably reactionary.

The political challenge for the postmodern left, as I see it, demands first a recognition and cogent interpretation of the dramatic and often confusing fourth modernization of capitalism that is presently taking place. It is becoming increasingly clear that this profound restructuring cannot be practically and politically understood only with the conventional tools and insights of modern Marxism or radical social science. This does not mean that these tools and insights need to be abandoned, as many formerly on the modern left have rushed to do. They must instead be flexibly and adaptively restructured to contend more effectively with a contemporary capitalism that is itself becoming more flexibly and adaptively reconstituted. The reactionary postmodern politics of Reaganism and Thatcherism, for example, must be directly confronted with an informed postmodern politics of resistance and demystification, one that can pull away the deceptive ideological veils that are today reifying and obscuring, in new and different ways, the restructured instrumentalities of class exploitation, gender and racial domination, cultural and personal disempowerment, and environmental degradation. The debates on the perils and possibilities of postmodernity must be joined, not abandoned, for the making of both history and geography is at stake.

I do not propose to construct a radical postmodern political programme here. But I do want to make sure that the project, however it takes shape, is consciously spatialized from the outset. We must be insistently aware of how space can be made to hide consequences from us, how relations of power and discipline are inscribed into the apparently innocent spatiality of social life, how human geographies become filled with politics and ideology. Every one of the nine essays can accordingly be read as an attempted spatialization, a post-scripted effort to compose a new critical human geography, an historical and geographical materialism attuned to the contemporary political and theoretical challenges.

A direct critique of historicism – without stumbling into a simplistic anti-history – is a necessary step forward in this spatialization of critical thought and political action. The first four essays reverse the imposing tapestry of historicism to trace the submergence and eventual reassertion of space in critical social theory through the evolving encounter between the disciplines and discourses of Western Marxism and Modern Geography. The distinctively Marxist geography that eventually arose from this encounter, and the French Marxisms that so influentially shaped the theoretical debates, are given particular attention, for they nurtured almost alone a critical discourse in which space 'mattered', in which human geography was not entirely subsumed within the historical imagination.

In chapters 3 and 4, I look back to my earlier writings on the socio-spatial dialectic, the theoretical specificity of the urban, and the vital role of geographically uneven development in the survival of capitalism. These three themes have provided important springboards for the reassertion of space in critical social theory via the spatialization of fundamental Marxist concepts and modes of analysis. Standing alone, however, these chapters may have a somewhat hollow ring, for they depend almost entirely on logical persuasion and assertive theoretical argument, couched in the rhetorical language of a fairly conventional Marxism. The last three essays attempt to give greater empirical and interpretive substance to these arguments, while the first two help to explain their historical origins and development. In chapters 5 and 6, however, I take another path of reinforcement and demonstration that delves into the 'backward linkages' from theoretical argument to the more abstract realm of ontology. In many ways, these middle chapters are pivotal for the entire collection of essays. They too can be read first to provide a different introduction.

The reassertion of space and the interpretation of postmodern geographies are not only a focus for empirical investigation, responding to a call for increasing attention to spatial form in concrete social research and informed political practice. Neither is the reassertion of space simply a metaphorical recomposition of social theory, a superficial linguistic spatialization that makes geography appear to matter theoretically as much as history. Taking space seriously requires a much deeper deconstruction and reconstitution of critical thought and analysis at every level of abstraction, including ontology. Especially ontology, perhaps, for it is at this fundamental level of existential discussion that the despatializing distortions of historicism are most firmly anchored.

Chapter 5 opens the ontological deconstruction with some observations by a spatially re-awakened Nicos Poulantzas, echoing both Lefebvre and Foucault, on the illusions of space and time that have characterized the history of Western Marxism. Of particular significance is Poulantzas's conceptualization of the spatial 'matrix' of the state and society as simultaneously the presupposition and embodiment of the relations of production, a 'primal material framework' rather than merely a mode of 'representation'. I take these observations further to argue that two persistent illusions have so dominated Western ways of seeing space that they have blocked from critical interrogation a third interpretive geography, one which recognizes spatiality as simultaneously (there's that word again) a social product (or outcome) and a shaping force (or medium) in social life: the crucial insight for both the socio-spatial dialectic and an historico-geographical materialism.

The 'illusion of opaqueness' reifies space, inducing a myopia that sees only a superficial materiality, concretized forms susceptible to little else but measurement and phenomenal description: fixed, dead, and undialectical: the Cartesian cartography of spatial science. Alternatively, the 'illusion of transparency' dematerializes space into pure ideation and representation, an intuitive way of thinking that equally prevents us from seeing the social construction of affective geographies, the concretization of social relations embedded in spatiality, an interpretation of space as a 'concrete abstraction', a social hieroglyphic similar to Marx's conceptualization of the commodity form. Philosophers and geographers have tended to bounce back and forth between these two distorting illusions for centuries, dualistically obscuring from view the power-filled and problematic making of geographies, the enveloping and instrumental spatialization of society.

Breaking through this double bind involves an ontological struggle to restore the meaningful existential spatiality of being and human consciousness, to compose a social ontology in which space matters from the very beginning. I engage in this struggle first in a critical re-evaluation of the time-warped ontologies of Sartre and Heidegger, the two most influential theorists of being in the twentieth century; and then, in chapter 6, in an analysis and extension of the revamped social ontology of 'time-space structuration' being developed by Anthony Giddens. Building on Giddens, one can see more clearly an existentially structured spatial topology and *topos* attached to being-in-the-world, a primordial contextualization of social being in a multi-layered geography of socially created and differentiated nodal regions nesting at many different scales around the mobile personal spaces of the human body and the more fixed communal locales of human settlements. This ontological spatiality situates the human subject in a formative geography once and for all, and provokes the need for a radical reconceptualization of epistemology, theory construction, and empirical analysis.

The construction of a spatialized ontology is as much a voyage of geographical exploration and discovery as are the essays on Los Angeles or the attempts to reveal the critical silences of historicism. It helps to complete an introductory and indicative map for the collection of essays, defining their scope, charting their interpretive terrain, and identifying some of the paths to be travelled. The composite picture is still incomplete, for there is so much yet to be discovered and explored in the contemporary reassertion of space in social theory, so much further to go before we can be sure of the impact and implications of *Postmodern Geographies*.

Despite the playful ploy of combining a postscript with a preface, I will refrain from making additional qualifying and self-critical reflections on the finished product and conclude instead with some necessary acknowledgements. First, I must express my debt to those I have chosen to present as the pioneers of postmodern geographies: Michel Foucault, John Berger, Ernest Mandel, Fredric Jameson, Marshall Berman, Nicos Poulantzas, Anthony Giddens, David Harvey, and especially Henri Lefebvre, whose insistent and inspiring sense of spatiality made me feel less alone over the past decade. These figures have never described themselves as postmodern geographers, but I believe they are and try to explain why by selectively appropriating their insights.

Since every one of the nine essays draws in part on my already published work, I must also thank the publishers and editors of the following books and journals for allowing me to excerpt and rewrite freely: *New Models in Geography* (R. Peet and N. Thrift, eds), Allen and Unwin (for sections of chapters 1 and 2); *Annals of the Association of American Geographers* and *Antipode* (for most of chapters 3 and 4); *Social Relations and Spatial Structures* (D. Gregory and J. Urry, eds), Macmillan Education Ltd and St Martin's Press (the first half of chapter 5); *Environment and Planning A* (the first half of chapter 6); *Environment and Planning D: Society and Space* (the regional half of chapter 7, and almost all of chapter 9); *Economic Geography* and *The Capitalist City* (M. Smith and J. Feagin, eds), Basil Blackwell (for much of chapter 8). The full references can be found in the bibliography.

I have only the University of California, Los Angeles, to thank for funding my research and writing over the past ten years — and even more so for sustaining a remarkably stimulating academic environment within and around the Graduate School of Architecture and Urban Planning. My colleagues and students in the Urban Planning Program have been especially supportive and tolerant, dragging me back to a practical reality whenever I ventured too far into the realms of spatial abstraction. Costis Hadjimichalis, Rebecca Morales, Goetz Wolff, Allan Heskin, Marco Cenzatti, and Allen Scott have co-authored some of the articles I have selectively recomposed in this collection of essays; and I have benefitted immensely from the editorial acumen, stimulating ideas, and encouraging words of Mike Davis and Margaret FitzSimmons. I am grateful for all their contributions.

Finally, I must acknowledge Maureen, the most practical, insistent and sustaining realist of them all. No one is more pleased that this book is finished.

# 1

# History:
# Geography:Modernity

Did it start with Bergson or before? Space was treated as the dead, the
fixed, the undialectical, the immobile. Time, on the contrary was richness,
fecundity, life, dialectic. (Foucault, 1980, 70)

The great obsession of the nineteenth century was, as we know, history:
with its themes of development and of suspension, of crisis and cycle,
themes of the ever-accumulating past, with its great preponderance of dead
men and the menacing glaciation of the world.... The present epoch will
perhaps be above all the epoch of space. We are in the epoch of simultaneity: we are in the epoch of juxtaposition, the epoch of the near and far,
of the side-by-side, of the dispersed. We are at a moment, I believe, when
our experience of the world is less that of a long life developing through
time than that of a network that connects points and intersects with its own
skein. One could perhaps say that certain ideological conflicts animating
present-day polemics oppose the pious descendants of time and the determined inhabitants of space. (Foucault, 1986, 22)

The nineteenth-century obsession with history, as Foucault described it,
did not die in the *fin de siècle*. Nor has it been fully replaced by a spatialization of thought and experience. An essentially historical epistemology
continues to pervade the critical consciousness of modern social theory.
It still comprehends the world primarily through the dynamics arising
from the emplacement of social being and becoming in the interpretive contexts of time: in what Kant called *nacheinander* and Marx
defined so transfiguratively as the contingently constrained 'making of
history'. This enduring epistemological presence has preserved a privileged place for the 'historical imagination' in defining the very nature of
critical insight and interpretation.

So unbudgeably hegemonic has been this historicism of theoretical

*10*

consciousness that it has tended to occlude a comparable critical sensibility to the spatiality of social life, a practical theoretical consciousness that sees the lifeworld of being creatively located not only in the making of history but also in the construction of human geographies, the social production of space and the restless formation and reformation of geographical landscapes: social being actively emplaced in space *and* time in an explicitly historical *and* geographical contextualization. Although others joined Foucault to urge a rebalancing of this prioritization of time over space, no hegemonic shift has yet occurred to allow the critical eye – or the critical I – to see spatiality with the same acute depth of vision that comes with a focus on *durée.* The critical hermeneutic is still enveloped in a temporal master-narrative, in a historical but not yet comparably geographical imagination. Foucault's revealing glance back over the past hundred years thus continues to apply today. Space still tends to be treated as fixed, dead, undialectical; time as richness, life, dialectic, the revealing context for critical social theorization.

As we move closer to the end of the twentieth century, however, Foucault's premonitory observations on the emergence of an 'epoch of space' assume a more reasonable cast. The material and intellectual contexts of modern critical social theory have begun to shift dramatically. In the 1980s, the hoary traditions of a space-blinkered historicism are being challenged with unprecedented explicitness by convergent calls for a far-reaching spatialization of the critical imagination. A distinctively postmodern and critical human geography is taking shape, brashly reasserting the interpretive significance of space in the historically privileged confines of contemporary critical thought. Geography may not yet have displaced history at the heart of contemporary theory and criticism, but there is a new animating polemic on the theoretical and political agenda, one which rings with significantly different ways of seeing time and space together, the interplay of history and geography, the 'vertical' and 'horizontal' dimensions of being in the world freed from the imposition of inherent categorical privilege.

It remains all too easy for even the best of the 'pious descendants of time' to respond to these pesky postmodern intrusions with an antidisestablishmentarian wave of a still confident upper hand or with the presumptive yawns of a seen-it-all-before complacency. In response, the determined intruders often tend to overstate their case, creating the unproductive aura of an anti-history, inflexibly exaggerating the critical privilege of contemporary spatiality in isolation from an increasingly silenced embrace of time. But from these confrontational polemics is also arising something else, a more flexible and balanced critical theory that re-entwines the making of history with the social production of space, with the construction and configuration of human geographies.

New possibilities are being generated from this creative commingling, possibilities for a simultaneously historical and geographical material- ism; a triple dialectic of space, time, and social being; a transformative re-theorization of the relations between history, geography, and modernity.

We are not yet sure enough about this incipient spatialization of critical theory to give a comprehensive and confident epistemological account; too much is at stake to attempt a premature totalization of a still shifting discourse. Nevertheless, the development of what I call postmodern geographies has progressed far enough to have changed significantly both the material landscape of the contemporary world and the interpretive terrain of critical theory. The time has come, then, for at least a first round of responsive evaluation of these two changing contexts of history and geography, modernity and postmodernity – one imprinted concretely on the empirical fabric of contemporary life (a postmodern geography of the material world), and the other threading through the ways we make practical and political sense of the present, the past, and the potential future (a postmodern geography of critical social consciousness).

In this opening chapter I will trace a reconfigurative path through the intellectual history of critical social theory from the last *fin de siècle* to the present, picking out the hidden narrative that has instigated the contemporary reassertion of space. My intent is not to erase the histori- cal hermeneutic but to open up and recompose the territory of the historical imagination through a critical spatialization. As will be evident in each subsequent chapter, this reassertion of space in critical social theory is an exercise in both deconstruction and reconstitution. It cannot be accomplished simply by appending spatial highlights to inherited critical perspectives and sitting back to watch them glow with logical conviction. The stranglehold of a still addictive historicism must first be loosened. The narrative task is effectively described by Terry Eagleton in *Against the Grain* (1986, 80):

> To 'deconstruct', then, is to reinscribe and resituate meanings, events and objects within broader movements and structures; it is, so to speak, to reverse the imposing tapestry in order to expose in all its unglamorously dishevelled tangle the threads constituting the well heeled image it presents to the world.

## Locating the Origins of Postmodern Geographies

The first insistent voices of postmodern critical human geography appeared in the late 1960s, but they were barely heard against the then prevailing temporal din. For more than a decade, the spatializing project remained strangely muted by the untroubled reaffirmation of the primacy of history over geography that enveloped both Western Marxism and liberal social science in a virtually sanctified vision of the ever-accumulating past. One of the most comprehensive and convincing pictures of this continuously historical contextualization was drawn by C. Wright Mills in his paradigmatic portrayal of the sociological imagination (Mills, 1959). Mills's work provides a useful point of departure for spatializing the historical narrative and reinterpreting the course of critical social theory.

## The silenced spatiality of historicism

Mills maps out a sociological imagination that is deeply rooted in an historical rationality – what Martin Jay (1984) would call a 'longitudinal totalization' – that applies equally well to critical social science and to the critical traditions of Marxism.

[The sociological imagination] is a quality of mind that will help [individuals] to use information and to develop reason in order to achieve lucid summations of what is going on in the world and of what may be happening within themselves. (1959, 11)

The first fruit of this imagination – and the first lessons of the social science that embodies it – is the idea that the individual can understand his own experience and gauge his own fate only by locating himself within his period, that he can know his own chances in life only by becoming aware of those of all individuals in his circumstances.... We have come to know that every individual lives, from one generation to the next, in some society; that he lives out a biography, and that he lives it out within some historical sequence. By the fact of his living he contributes, however minutely, to the shaping of this society and to the course of history, even as he is made by society and by its historical push and shove. (12)

He goes further:

The sociological imagination enables us to grasp history and biography and the relations of the two within society. This is its task and its promise. To recognize this task and this promise is the mark of the classic social analyst.... *No social study that does not come back to the problems of biography, of history, and of their intersections within society, has completed its intellectual journey.* (Ibid., emphasis added)

I draw upon Mills's depiction of what is essentially a historical imagination to illustrate the alluring logic of historicism, the rational reduction of meaning and action to the temporal constitution and experience of social being. This connection between the historical imagination and historicism needs further elaboration. First, there is the easier question of why 'sociological' has been changed to 'historical'. As Mills himself notes, 'every cobbler thinks leather is the only thing', and as a trained sociologist Mills names his leather after his own disciplinary specialization and socialization. The nominal choice personally specifies what is a much more widely shared 'quality of mind' that Mills claims should pervade, indeed embody, all social theory and analysis, an emancipatory rationality grounded in the intersections of history, biography, and society.

To be sure, these 'life-stories' have a geography too; they have milieux, immediate locales, provocative emplacements which affect thought and action. The historical imagination is never completely spaceless and critical social historians have written, and continue to write, some of the best geographies of the past. But it is always time and history that provide the primary 'variable containers' in these geographies. This would be just as clear whether the critical orientation is described as sociological or political or anthropological – or for that matter phenomenological, existential, hermeneutic, or historical materialist. The particular emphases may differ, but the encompassing perspective is shared. An already-made geography sets the stage, while the wilful making of history dictates the action and defines the story line.

It is important to stress that this historical imagination has been particularly central to critical social theory, to the search for practical understanding of the world as a means of emancipation versus maintenance of the status quo. Social theories which merely rationalize existing conditions and thereby serve to promote repetitive behaviour, the continuous reproduction of established social practices, do not fit the definition of critical theory. They may be no less accurate with respect to what they are describing, but their rationality (or irrationality, for that matter) is likely to be mechanical, normative, scientific, or instrumental rather than critical. It is precisely the critical and potentially emancipatory value of the historical imagination, of people 'making history' rather than taking it for granted, that has made it so compulsively appealing. The constant reaffirmation that the world can be changed by human action, by praxis, has always been the centrepiece of critical social theory whatever its particularized source and emphasis.

The development of critical social theory has revolved around the assertion of a mutable history against perspectives and practices that mystify the changeability of the world. The critical historical discourse thus sets itself against abstract and transhistorical universalizations (including notions of a general 'human nature' which explain everything and nothing at the same time); against naturalisms, empiricisms, and positivisms which proclaim physical determinations of history apart from social origins; against religious and ideological fatalisms which project spiritual determinations and teleologies (even when carried forward in the cloak of human consciousness); against any and all conceptualizations of the world which freeze the frangibility of time, the possibility of 'breaking' and remaking history.

Both the attractive critical insight of the historical imagination and its continuing need to be forcefully defended against distracting mystifications have contributed to its exaggerated assertion as historicism. Historicism has been conventionally defined in several different ways. Raymond Williams' *Keywords* (1983), for example, presents three contemporary choices, which he describes as: 1) 'neutral' – a method of study using facts from the past to trace the precedents of current events; 2) 'deliberate' – an emphasis on variable historical conditions and contexts as a privileged framework for interpreting all specific events; and 3) 'hostile' – an attack on all interpretation and prediction which is based on notions of historical necessity or general laws of historical development.

I wish to give an additional twist to these options by defining historicism as an overdeveloped historical contextualization of social life and social theory that actively submerges and peripheralizes the geographical or spatial imagination. This definition does not deny the extraordinary power and importance of historiography as a mode of emancipatory insight, but identifies historicism with the creation of a critical silence, an implicit subordination of space to time that obscures geographical interpretations of the changeability of the social world and intrudes upon every level of theoretical discourse, from the most abstract ontological concepts of being to the most detailed explanations of empirical events.

This definition may appear rather odd when set against the long tradition of debate over historicism that has flourished for centuries.[1] The failure of this debate to recognize the peculiar theoretical peripheralization of space that has accompanied even the most neutral forms of historicism is, however, precisely what began to be discovered in the late 1960s, in the ragged beginnings of what I have called a postmodern

---

1. See Popper (1957), Eliade (1959), Lowith (1949), Cohen (1978) and Rorty (1980), for a sampling of very different approaches to historicism. Rorty, in *Philosophy and the Mirror of Nature* (9), makes the interesting comment that traditional Cartesian–Kantian philosophy was 'an attempt to escape from history ... to find nonhistorical conditions of any possible historical development'. The key figures of twentieth-century analytical philosophy espoused by Rorty – Wittgenstein, Dewey, and Heidegger – are then presented as restoratively historicist. Rorty adds: 'The moral of this book is also historicist' (10). Human geography characteristically disappears almost entirely in this modern mirroring of nature, except as an archaic reflection.

critical human geography. Even then, the main currents of critical social thought had become so spatially-blinkered that the most forceful re-assertions of space versus time, geography versus history, had little effect. The academic discipline of Modern Geography had, by that time, been rendered theoretically inert and contributed little to these first reassertions. And when some of the most influential social critics of the time took a bold spatial turn, not only was it usually seen by the un-converted as something else entirely, but the turners themselves often chose to muffle their critiques of historicism in order to be understood at all.

Only a few particularly vigorous voices resonated through the still hegemonic historicism of the past twenty years to pioneer the development of postmodern geography. The most persistent, insistent, and consistent of these spatializing voices belonged to the French Marxist philosopher, Henri Lefebvre. His critical theorization of the social production of space will thread its way into every subsequent chapter. Here, however, I will extract and represent the spatializing projects of two other critical theorists, Michel Foucault and John Berger, whose assertive postmodern geographies have been largely hidden from view by their more comforting and familiar identification as historians.

### The ambivalent spatiality of Michel Foucault

The contributions of Foucault to the development of critical human geography must be drawn out archeologically, for he buried his pre-cursory spatial turn in brilliant whirls of historical insight. He would no doubt have resisted being called a postmodern geographer, but he was one, *malgré lui,* from *Madness and Civilization* (1961) to his last works on *The History of Sexuality* (1978). His most explicit and revealing observations on the relative significance of space and time, however, appear not in his major published works but almost innocuously in his lectures and, after some coaxing interrogation, in two revealing inter-views: 'Questions on Geography' (Foucault, 1980) and 'Space, Know-ledge, and Power' (Rabinow, 1984; see also Wright and Rabinow, 1982).

The epochal observations which head this chapter, for example, were first made in a 1967 lecture entitled '*Des Espaces Autres*'. They remained virtually unseen and unheard for nearly twenty years, until their publication in the French journal *Architecture-Mouvement-Continuité* in 1984 and, translated by Jay Miskowiec as 'Of Other Spaces', in *Diacritics* (1986). In these lecture notes, Foucault outlined his notion of 'heterotopias' as the characteristic spaces of the modern world, superseding the hierarchic 'ensemble of places' of the Middle

Ages and the enveloping 'space of emplacement' opened up by Galileo into an early–modern, infinitely unfolding, 'space of extension' and measurement. Moving away from both the 'internal space' of Bachelard's brilliant poetics (1969) and the intentional regional descriptions of the phenomenologists, Foucault focused our attention on another spatiality of social life, an 'external space', the actually lived (and socially produced) space of sites and the relations between them:

> The space in which we live, which draws us out of ourselves, in which the erosion of our lives, our time and our history occurs, the space that claws and gnaws at us, is also, in itself, a heterogeneous space. In other words, we do not live in a kind of void, inside of which we could place individuals and things. We do not live inside a void that could be colored with diverse shades of light, we live inside a set of relations that delineates sites which are irreducible to one another and absolutely not superimposable on one another. (1986, 23)

These heterogeneous spaces of sites and relations – Foucault's heterotopias – are constituted in every society but take quite varied forms and change over time, as 'history unfolds' in its adherent spatiality. He identifies many such sites: the cemetery and the church, the theatre and the garden, the museum and the library, the fairground and the 'vacation village', the barracks and the prison, the Moslem hammam and the Scandinavian sauna, the brothel and the colony. Foucault contrasts these 'real places' with the 'fundamentally unreal spaces' of utopias, which present society in either 'a perfected form' or else 'turned upside down':

> The heterotopia is capable of juxtaposing in a single real place several spaces, several sites that are in themselves incompatible ... they have a function in relation to all the space that remains. This function unfolds between two extreme poles. Either their role is to create a space of illusion that exposes every real space, all the sites inside of which human life is partitioned, as still more illusory.... Or else, on the contrary, their role is to create a space that is other, another real space, as perfect, as meticulous, as well arranged as ours is messy, ill constructed, and jumbled. The latter type would be the heterotopia, not of illusion, but of compensation, and I wonder if certain colonies have not functioned somewhat in this manner. (1986, 25, 27)

With these remarks, Foucault exposed many of the compelling directions he would take in his lifework and indirectly raised a powerful argument against historicism – and against the prevailing treatments of space in the human sciences. Foucault's heterogeneous and relational space of heterotopias is neither a substanceless void to be filled by cognitive intuition nor a repository of physical forms to be phenomenologically described in all its resplendent variability. It is another space, what

*concrete & abstract at same time*
*habitus of soc practices*

Lefebvre would describe as *l'espace vécu,* actually lived and socially created spatiality, concrete and abstract at the same time, the habitus of social practices. It is a space rarely seen for it has been obscured by a bi-focal vision that traditionally views space as either a mental construct or a physical form – a dual illusion that I discuss in greater detail in chapter 5.

To illustrate his innovative interpretation of space and time and to clarify some of the often confusing polemics which were arising around it, Foucault turned to the then current debates on structuralism, one of the twentieth-century's most important avenues for the reassertion of space in critical social theory. Foucault vigorously insisted that he himself was not (just?) a structuralist, but he recognized in the development of structuralism a different and compelling vision of history and geography, a critical reorientation that was connecting space and time in new and revealing ways.

> Structuralism, or at least that which is grouped under this slightly too general name, is the effort to establish, between elements that could have been connected on a temporal axis, an ensemble of relations that makes them appear as juxtaposed, set off against one another, in short, as a sort of configuration. Actually structuralism does not entail a denial of time; it does involve a certain manner of dealing with what we call time and what we call history. (1986, 22)

This synchronic 'configuration' is the spatialization of history, the making of history entwined with the social production of space, the structuring of a historical geography.[2]

Foucault refused to project his spatialization as an anti-history but his history was provocatively spatialized from the very start. This was not just a shift in metaphorical preference, as it frequently seemed to be for Althusser and others more comfortable with the structuralist label than Foucault. It was the opening up of history to an interpretive geography. To emphasize the centrality of space to the critical eye, especially regarding the contemporary moment, Foucault becomes most explicit:

> In any case I believe that the anxiety of our era has to do fundamentally with space, no doubt a great deal more than with time. Time probably appears to

---

2. Structuralism's presumed 'denial' of history has triggered an almost maniacal attack on its major proponents by those imbued most rigidly with an emancipatory historicism. What Foucault is suggesting, however, is that structuralism is not an anti-history but an attempt to deal with history in a different way, as a spatio-temporal configuration, simultaneously and interactively synchronic and diachronic (to use the conventional categorical opposition).

us only as one of the various distributive operations that are possible for the elements that are spread out in space. (Ibid., 23)

He would never be quite so explicit again. Foucault's spatialization took on a more demonstrative rather than declarative stance, confident perhaps that at least the French would understand the intent and significance of his strikingly spatialized historiography.

In an interview conducted shortly before his death (Rabinow, 1984), Foucault reminisced on his exploration 'Of Other Spaces' and the enraged reactions it engendered from those he once identified as the 'pious descendants of time'. Asked whether space was central to the analysis of power, he answered:

Yes. Space is fundamental in any form of communal life; space is fundamental in any exercise of power. To make a parenthetical remark, I recall having been invited, in 1966, by a group of architects to do a study of space, of something that I called at that time 'heterotopias', those singular spaces to be found in some given social spaces whose functions are different or even the opposite of others. The architects worked on this, and at the end of the study someone spoke up – a Sartrean psychologist – who firebombed me, saying that *space* is reactionary and capitalist, but *history* and *becoming* are revolutionary. This absurd discourse was not at all unusual at the time. Today everyone would be convulsed with laughter at such a pronouncement, but not then.

Amidst today's laughter – still not as widespread and convulsive as Foucault assumed it would be – one can look back and see that Foucault persistently explored what he called the 'fatal intersection of time with space' from the first to the last of his writings. And he did so, we are only now beginning to realize, infused with the emerging perspective of a post-historicist and postmodern critical human geography.

Few could see Foucault's geography, however, for he never ceased to be a historian, never broke his allegiance to the master identity of modern critical thought. To be labelled a geographer was an intellectual curse, a demeaning association with an academic discipline so far removed from the grand houses of modern social theory and philosophy as to appear beyond the pale of critical relevance. Foucault had to be coaxed into recognizing his formative attachment to the geographer's spatial perspective, to admit that geography was always at the heart of his concerns. This retrospective admission appeared in an interview with the editors of the French journal of radical geography, *Herodote*, and was published in English as 'Questions on Geography', in *Power/ Knowledge* (Foucault, 1980). In this interview, Foucault expanded upon the observations he made in 1967, but only after being pushed to do so by the interviewers.

At first, Foucault was surprised – and annoyed – at being asked by his interviewers why he had been so silent about the importance of geography and spatiality in his works despite the profuse use of geographical and spatial metaphors. The interviewers suggested to him:

> If geography is invisible or ungrasped in the area of your explorations and excavations, this may be due to the deliberately historical or archeological approach which privileges the factor of time. Thus one finds in your work a rigorous concern with periodization that contrasts with the vagueness of your spatial demarcations.

Foucault responded immediately by diversion and inversion, throwing back the responsibility for geography to his interviewers (while remembering the critics who reproached him for his 'metaphorical obsession' with space). After further questioning, however, he admitted (again?) that space has been devalued for generations by philosophers and social critics, reasserted the inherent spatiality of power/knowledge, and ended with a *volte face*:

> I have enjoyed this discussion with you because I've changed my mind since we started. I must admit that I thought you were demanding a place for geography like those teachers who protest when an education reform is proposed because the number of hours of natural sciences or music is being cut.... Now I can see that the problems you put to me about geography are crucial ones for me. Geography acted as the support, the condition of possibility for the passage between a series of factors I tried to relate. Where geography itself was concerned, I either left the question hanging or established a series of arbitrary connections.... Geography must indeed lie at the heart of my concerns. (Foucault, 1980, 77)

Foucault's argument here takes a new turn, from simply looking at 'other spaces' to questioning the origins of 'this devaluation of space that has prevailed for generations'. It is at this point that he makes the comment cited earlier on the post-Bergsonian treatment of space as passive and lifeless, time as richness, fecundity, dialectic.

Here then are the inquisitive ingredients for a direct attack on historicism as the source of the devaluation of space, but Foucault had other things in mind. In a revealing aside, he takes an integrative rather than deconstructive path, holding on to his history but adding to it the crucial nexus that would flow through all his work: the linkage between space, knowledge, and power.

> For all those who confuse history with the old schemas of evolution, living continuity, organic development, the progress of consciousness or the project

of existence, the use of spatial terms seems to have an air of an anti-history. If one started to talk in terms of space that meant one was hostile to time. It meant, as the fools say, that one 'denied history', that one was a 'technocrat'. They didn't understand that to trace the forms of implantation, delimitation and demarcation of objects, the modes of tabulation, the organisation of domains meant the throwing into relief of processes – historical ones, needless to say – of power. The spatialising description of discursive realities gives on to the analysis of related effects of power. (Ibid.)

In 'The Eye of Power', published as a preface to Jeremy Bentham, *La Panoptique* (1977) and reprinted in *Power/Knowledge* (Foucault, 1980, ed. Gordon, 149), he restates his ecumenical project:

A whole history remains to be written of *spaces* – which would at the same time be the history of *powers* (both of these terms in the plural) – from the great strategies of geopolitics to the little tactics of the habitat.

Foucault thus postpones a direct critique of historicism with an acute lateral glance, at once maintaining his spatializing project but preserving his historical stance. 'History will protect us from historicism', he optimistically concludes (Rabinow, 1984, 250).

I will return to Foucault's provocative spatialization of power in later chapters. For now, I have used his work to illustrate one almost invisible but nevertheless formative career in postmodern critical human geography, a career hidden from explicit recognition as geographical by the persistent hegemony of historicism. Another similarly hidden (historical) geography can be found in the works of John Berger, one of the most influential and innovative art critics writing in English today.

### Envisioning space through the eyes of John Berger

Like Foucault, John Berger dwells on the intersection of time and space in virtually all his writings. Amongst his most recent works is a play entitled *A Question of Geography* and a personalized volume of poetry and prose that conceives visually of love, *And our faces, my heart, as brief as photos* (Berger, 1984). Symbolizing his insistent balancing of history and geography, lineage and landscape, period and region, Berger opens this slim volume by stating 'Part One is About Time. Part Two is About Space'. The embracing themes follow accordingly: the first part labelled 'Once', the second 'Here': neither one inherently privileged, both necessarily faceted together. But Berger does make an explicit choice in at least one of his earlier writings and it is upon this assertive choice that I wish to focus attention.

In what still stands today as perhaps the most direct declaration of the end of historicism, this most spatially visionary of art historians – dare one call him an art geographer? – calls openly for a spatialization of critical thought. In the following passage, from *The Look of Things* (1974, 40), Berger condenses the essence of postmodern geographies in a spatially politicized aesthetic:

> We hear a lot about the crisis of the modern novel. What this involves, fundamentally, is a change in the *mode of narration*. It is scarcely any longer possible to tell a straight story sequentially unfolding in time. And this is because we are too aware of what is continually traversing the storyline *laterally*. That is to say, instead of being aware of a point as an infinitely small part of a straight line, we are aware of it as an infinitely small part of an infinite number of lines, as the centre of a star of lines. Such awareness is the result of our constantly having to take into account the *simultaneity and extension* of events and possibilities.
>
> There are so many reasons why this should be so: the range of modern means of communication: the scale of modern power: the degree of personal political responsibility that must be accepted for events all over the world: the fact that the world has become indivisible: the unevenness of economic development within that world: the scale of the exploitation. All these play a part. *Prophesy now involves a geographical rather than historical projection; it is space not time that hides consequences from us.* To prophesy today it is only necessary to know men [and women] as they are throughout the whole world in all their inequality. Any contemporary narrative which ignores the urgency of this dimension is incomplete and acquires the oversimplified character of a fable (emphases and brackets mine).

This pointed passage pops out of an essay on modern portrait painting in which Berger tries to explain why the historical significance of portraiture, so often in the past the visual personification of authoritative lineage and social (class) position, has changed so dramatically in the twentieth century. To make his point, he turns to an analogous change in the modern novel, a shift in the context of meaning and interpretation which hinges around the impress of simultaneity versus sequence, spatiality versus historicity, geography versus biography. In so doing, he begins to set into place a train of arguments that define the postmodern turn against historical determinations and vividly announce the need for an explicitly spatialized narrative.

The first of these assertively postmodern geographical arguments rests on the recognition of a profound and crisis-induced restructuring of contemporary life, resulting in significant changes in 'the look of things' and, if I may continue to draw upon Berger's captivating book titles, in our 'ways of seeing' (1972). This restructuring, for Berger, involves a fundamental recomposition of the 'mode of narration', arising from a new awareness that we must take into account 'the simultaneity and extension of events and possibilities' to make sense of what we see. We can no longer depend on a story-line unfolding sequentially, an ever-accumulating history marching straight forward in plot and denouement, for too much is happening against the grain of time, too much is continually traversing the story-line laterally. A contemporary portrait no longer directs our eye to an authoritative lineage, to evocations of heritage and tradition alone. Simultaneities intervene, extending our point of view outward in an infinite number of lines connecting the subject to a whole world of comparable instances, complicating the temporal flow of meaning, short-circuiting the fabulous stringing-out of 'one damned thing after another'. The new, the novel, now must involve an explicitly geographical as well as historical configuration and projection.

To explain why this is so, Berger astutely situates the restructured narrative in a pervasive context and consciousness of geographically uneven development, into a constellation of lines and photography of surfaces connecting every (hi)story to an attention-shaping horizontality that stretches everywhere in its power, indivisibility, exploitation, and inequality. Our urgent awareness of geographically uneven development and the revived sense of our personal political responsibility for it as a product we have collectively created, spatializes the contemporary moment and reveals the insights to be derived from a deeper understanding of contemporary crisis and restructuring in literature and science, in our daily lives and in the conditions of men and women 'as they are throughout the whole world in all their inequality'. I repeat again Berger's provocative conclusion: Prophesy now involves a geographical rather than historical projection; it is space not time that hides consequences from us.

What a shattering assertion for those who see only through the spectacles of time. Arising from the recognition of a profound restructuring of contemporary life and an explicit consciousness of geographically (and not just historically) uneven development is an extraordinary call for a new critical perspective, a different way of seeing the world in which human geography not only 'matters' but provides the most revealing critical perspective.

Before jumping to other conclusions, however, let us not forget that this spatialization of critical thought does not have to project a simplistic anti-history. As with Foucault, the reassertion of space in critical social theory does not demand an antagonistic subordination of time and history, a facile substitution and replacement. It is instead a call for an appropriate interpretive balance between space, time, and social being, or what may now more explicitly be termed the creation of human geographies, the making of history, and the constitution of society. To claim that, in the contemporary context, it is space not time that hides consequences from us is thus both an implied recognition that history has hitherto been accepted as the privileged mode of critical disclosure and discourse, and an argument that this privileged position, insofar as it has blocked from view the critical significance of the spatiality of social life, is no longer apt. It is the dominance of a historicism of critical thought that is being challenged, not the importance of history. Almost as if he were turning Mills's sociological imagination upside down, Berger notes that any contemporary narrative which ignores the urgency of the spatial dimension is incomplete and acquires the oversimplified character of a fable.

Berger thus joins with Foucault in pushing us towards a significant and necessary restructuring of critical social thought, a recomposition which enables us to see more clearly the long–hidden instrumentality of human geographies, in particular the encompassing and encaging spatializations of social life that have been associated with the historical development of capitalism. Foucault's path took him primarily into the microspaces of power, discipline, and surveillance, into the carceral city, the asylum, the human body. Berger's path continues to open up new ways of seeing art and aesthetics, portraits and landscapes, painters and peasants, in the past (once) and in the present (here). To crystallize and expand these spatial fields of insight and to attach postmodern critical human geography even more forcefully and explicitly to the instrumental spatiality of capitalism, the historical narrative must be re-entered at a different place and scale.

## The Deconstruction and Reconstitution of Modernity

In *All That is Solid Melts Into Air: The Experience of Modernity*, (1982), Marshall Berman explores the multiple reconfigurations of social life that have characterized the historical geography of capitalism over the past four hundred years. At the heart of his interpretive outlook is a revealing periodization of changing concepts of modernity from the formative sixteenth–century clash between the 'Ancients' and the 'Moderns' to the contemporary debates that herald still another conceptual and social reconfiguration, another reconsideration of what it means to be modern. In this concatenation of modernities is a history of historicism that can now begin to be written from a postmodern geographical perspective.

Berman broadly defines modernity as 'a mode of vital experience', a collective sharing of a particularized sense of 'the self and others', of 'life's possibilities and perils'. In this definition, there is a special place given to the ways we think about and experience time and space, history and geography, sequence and simultaneity, event and locality, the immediate period and region in which we live. Modernity is thus comprised of both context and conjuncture. It can be understood as the specificity of being alive, in the world, at a particular time and place; a vital individual and collective sense of contemporaneity. As such, the experience of modernity captures a broad mesh of sensibilities that reflects the specific and changing meanings of the three most basic and formative dimensions of human existence: space, time, and being. Herein lies its particular usefulness as a means of resituating the debates on history and geography in critical social theory and for defining the context and conjuncture of postmodernity.

Just as space, time, and matter delineate and encompass the essential qualities of the physical world, spatiality, temporality, and social being can be seen as the abstract dimensions which together comprise all facets of human existence. More concretely specified, each of these abstract existential dimensions comes to life as a social construct which shapes empirical reality and is simultaneously shaped by it. Thus, the spatial order of human existence arises from the (social) production of space, the construction of human geographies that both reflect and configure being in the world. Similarly, the temporal order is concretized in the making of history, simultaneously constrained and constraining in an evolving dialectic that has been the ontological crux of Marxist thought for over a hundred years. To complete the necessary existential triad, the social order of being-in-the-world can be seen as revolving around the constitution of society, the production and reproduction of social relations, institutions, and practices. How this ontological nexus of space–time–being is conceptually specified and given particular meaning in the explanation of concrete events and occurrences is the generative source of all social theory, critical or otherwise. It provides an illuminating motif through which to view the interplay between history, geography, and modernity.

*Sequences of modernity, modernization and modernism*

In the experience of modernity, the ontological nexus of social theory becomes specifically and concretely composed in a changing 'culture of time and space', to borrow the felicitous phrase used by Stephen Kern (1983) to describe the profound reconfiguration of modernity that took place in the previous *fin de siècle*.

From around 1880 to the outbreak of World War I a series of sweeping changes in technology and culture created distinctive new modes of thinking about and experiencing time and space. Technological innovations including the telephone, wireless telegraph, x-ray, cinema, bicycle, automobile and airplane established the material foundation for this reorientation; independent cultural developments such as the stream of consciousness novel, psychoanalysis, Cubism, and the theory of relativity shaped consciousness directly. The result was a transformation of the dimensions of life and thought. (1983, 1–2)

During this expanded *fin de siècle*, from the aftermath of the defeat of the Paris Commune to the events which would lead up to the Russian Revolution (to choose somewhat different turning points), the world changed dramatically. Industrial capitalism survived its predicted demise through a radical social and spatial restructuring which both intensified (or deepened, as in the rise of corporate monopolies and mergers) and extensified (or widened, as in the global expansion of imperialism) its definitive production relations and divisions of labour. Accompanying the rise of this new political economy of capitalism was an altered culture of time and space, a restructured historical geography taking shape from the shattered remains of an older order and infused with ambitious new visions and designs for the future as the very nature and experience of modernity – what it meant to be modern – was significantly reconstituted. A similar reconstitution took place in the prevailing forms of social theorization, equally attuned to the changing nature of capitalist modernity. But before turning to this restructuring of social theory, there is more to be derived from Berman's conceptualization of modernity and the recognition of the parallelism between the past and present *fins de siècle*.

As so many have begun to see, both *fin de siècle* periods resonate with similarly transformative, but not necessarily revolutionary, socio-spatial processes. As occurred roughly a century ago, there is currently a complex and conflictual dialectic developing between urgent socio-economic modernization sparked by the system-wide crises affecting contemporary capitalist societies; and a responsive cultural and political modernism aimed at making sense of the material changes taking place in the world and gaining control over their future directions. Modernization and modernism interact under these conditions of intensified crisis and restructuring to create a shifting and conflictful social context in which everything seems to be 'pregnant with its contrary', in which all that was once assumed to be solid 'melts into air', a description Berman borrows from Marx and represents as an essential feature of the vital experience of modernity-in-transition.

Modernization can be directly linked to the many different 'objective' processes of structural change that have been associated with the ability of capitalism to develop and survive, to reproduce successfully its fundamental social relations of production and distinctive divisions of labour despite endogenous tendencies towards debilitating crisis. This defining association between modernization and the survival of capitalism is crucial, for all too often the analysts of modernity extract social change from its social origins in modes of production to 'stage' history in idealized evolutionary modellings. From these perspectives, change just seems to 'happen' in a lock-step march of modernity replacing tradition, a mechanical teleology of progress. Modernization is not entirely the product of some determinative inner logic of capitalism, but neither is it a rootless and ineluctable idealization of history.

Modernization, as I view it here, is a continuous process of societal restructuring that is periodically accelerated to produce a significant recomposition of space-time-being in their concrete forms, a change in the nature and experience of modernity that arises primarily from the historical and geographical dynamics of modes of production. For the past four hundred years, these dynamics have been predominantly capitalist, as has been the very nature and experience of modernity during that time. Modernization is, like all social processes, unevenly developed across time and space and thus inscribes quite different historical geographies across different regional social formations. But on occasion, in the ever-accumulating past, it has become systemically synchronic, affecting all predominantly capitalist societies simultaneously. This synchronization has punctuated the historical geography of capitalism since at least the early nineteenth century with an increasingly recognizable macro-rhythm, a wave-like periodicity of societal crisis and restructuring that we are only now beginning to understand in all its ramifications.

Perhaps the earliest of these prolonged periods of 'global' crisis and restructuring stretched through what Hobsbawm termed the 'age of revolution' and peaked in the turbulent years between 1830 and 1848–51. The following decades were a time of explosive capitalist expansion in industrial production, urban growth, and international trade, the florescence of a classical, competitive, entrepreneurial regime of capital accumulation and social regulation. During the last three decades of the nineteenth century, however, boom turned largely into bust for the then most advanced capitalist countries as the Long Depression, as it was called, accentuated the need for another urgent restructuring and modernization, a new 'fix' for a capitalism forever addicted to crisis.

The same rollercoaster sequence of crisis-induced restructuring, leading to an expansionary boom, and then to crisis and restructuring again, marked the first half of the twentieth century, with the Great Depression echoing the conflictful system-wide downturns of the past and initiating

the transition from one distinctive regime of accumulation to another.
And as it now seems increasingly clear, the last half of the twentieth
century has followed a similar broad trajectory, with a prolonged expansionary period after the Second World War and a still ongoing, crisis-
filled era of attempted modernization and restructuring taking us toward
the next *fin de siècle*. The rhythm has been an insistent one, marking
time into what might be described as at least four metamorphic modernizations of capitalism, from the 1830s to the present.

The most rigorous and revealing analyses of this crisis-laden macro-
rhythm in the historical geography of capitalism have been made by
Ernest Mandel (1976, 1978, 1980). Mandel is particularly effective in
connecting the periodicity of intensified modernization with a series of
geographical restructurings similarly characterized by the attempt to
restore the supportive conditions for profitable capitalist accumulation
and labour control. This Mandelian periodization/regionalization of the
modernization process plays an important role in the development and
interpretation of postmodern geographies.

Berman also insightfully describes the characteristic features shaping
these periodically intensified modernizations. He presents the following
menu of material forces which contribute to the restructuring of the
experience of modernity as a collective sense of the 'perils and possibilities' of the contemporary:

the industrialization of production, which transforms scientific knowledge
into technology, creates new environments and destroys old ones, speeds up
the whole tempo of life, generates new forms of corporate power and class
struggle;

immense demographic upheavals, severing millions of people from their
ancestral habitats, hurtling them halfway across the world into new lives;

rapid and often cataclysmic urban growth;

systems of mass communications, dynamic in their development, enveloping
and binding together the most diverse people and societies;

increasingly powerful national states, bureaucratically structured and operated,
constantly striving to expand their powers;

mass social movements of people and peoples, challenging their political and
economic rulers, striving to gain control over their lives;

finally, bearing and driving all these people and institutions along, an ever-
expanding, drastically fluctuating capitalist world market. (Berman, 1982, 16)

This awesome catalogue vividly outlines the destructive creativity so
closely associated with both the modernization and survival of capitalism

over the past two centuries. It is today being repeated, with variations, once more.

Restructuring and modernization punctuate not only the concrete history and geography of capitalist development but also mark the changing course of critical social theory. To make this connection between the political economy of the empirical world and the world of theory brings us to Berman's conceptualization of modernism. In its broadest sense, modernism is the cultural, ideological, reflective, and, I will add, theory-forming response to modernization. It encompasses a heterogeneous array of subjective visions and strategic action programmes in art, literature, science, philosophy, and political practice which are unleashed by the disintegration of an inherited, established order and the awareness of the projected possibilities and perils of a restructured contemporary moment or conjuncture. Modernism is, in essence, a 'reaction formation', a conjunctural social movement mobilized to face the challenging question of what now is to be done given that the context of the contemporary has significantly changed. It is thus the culture-shaping, programmatic, and situated consciousness of modernity.

Each era of accelerated modernization has been a fertile spawning ground for powerful new modernisms emerging in almost every field of discourse and creativity. Of particular interest to the present narrative are two 'modern movements' which emerged around the turn of the nineteenth century to define separate and competitive realms of critical social theorization, one centred in the Marxist tradition, the other in more naturalist and positivist social science. Like other modern movements of the *fin de siècle*, they arose initially as rebelliously creative avant-garde movements challenging their own inherited orthodoxies with a new sense of what is to be done. To the traditionalists of the time, encased in older structures and strictures of modernity, the avant-garde movements appeared to dwell in a different world, in an alternative modernity, a 'postmodern' identity in the sense that it was no longer confinable within inherited and established traditions.

A Leninist Marxism was one of the most successful modern movements through the turn of the century, a reinvigorating and avant-garde restructuring of historical materialism/scientific socialism in both theory and practice: a modernized Marxism that significantly changed the world. Along with many other successful modern movements, Marxism–Leninism consolidated its victories in that part of the world which it changed most significantly. It also steeled itself against the great mid-century crises of the Great Depression and the Second World War, and moved into the last half of the twentieth century so formidably entrenched that it could no longer be described as avant-garde. The new became old, the vanguard became old guard: hegemonic, rigid, establishmentarian.

This historical process of rigidification at the core of twentieth-century Marxism split the movement geographically and in its approaches to theory and practice. A more conservative and assiduously pragmatic 'Eastern' mainstream deflected theoretical criticism and innovation to what has come to be called Western Marxism, removed enough from the central orthodoxies of Marxism–Leninism to be distinctive but too close to represent an autonomous modern movement of its own.[3] It is from within this 'peripheral' current of Western Marxism that the reassertion of space and the critique of historicism eventually emerge.

Arising in part as a reaction to the restructuring of Marxism was the consolidation of the Western, Modern, or, from a Marxist standpoint, bourgeois social sciences, a much more compartmentalized and fragmented intellectual division of labour than that which came forward from the modernization of the Marxist tradition. There were, nonetheless, some remarkable similarities. Both arose from the intellectual, political, and institutional struggles that developed in the late nineteenth century as competitive reinterpretations of how best to theorize and induce progressive changes in the modern social order (just as Marx and Comte attempted to do in response to the ending of the earlier age of revolution, after the first systemic modernization of capitalism). The social sciences also developed an internal division between an increasingly orthodox and hegemonic core tradition based largely on an instrumental and increasingly positivist appropriation of natural science methods in social analysis and theorization; and a collation of critical variants constantly pressing against disciplinary rigidification, fragmentation and scientism. This critical social science shared with Western Marxism two additional features: an emancipatory interest in the power of human consciousness and social will to break through all exogenous constraints; and a critical inscription of this social power and potentially revolutionary subjectivity in the 'making of history', in historical modes of explanation and interpretation, confrontation and critique.[4]

---

3. Precisely defining the boundaries and topography of Western Marxism is still a controversial issue. My definition is roughly the same as Anderson's (1976) for the period up to the 1960s, but closer to Jay's (1984) since then – that is, it encompasses continental and Anglo-American leftist scholars as well as several self-proclaimed non-Marxists of particular prominence, including, as Anderson seems loath to do, Michel Foucault.
4. Hughes (1958) still offers one of the best descriptions of the formation of this social science countertradition around the turn of the century. Twentieth–century critical social science has also mimicked the critical theories of Western Marxism in being too close to the mainstream to be defined as a separate modern movement but distinct enough to establish its own identifiable boundaries, traditions, and topography.

The *fin de siècle* thus brought with it a recomposed culture of time and space and a bifurcated critical social theory that was imbued, in both its major variants, with an invigorated historical imagination. What our accumulated histories of theoretical consciousness do not tell us, however, is that the modernisms which celebrated this historical imagination simultaneously induced a growing submergence and dissipation of the geographical imagination, a virtual annihilation of space by time in critical social thought and discourse. The quiet triumph of historicism has, in turn, deeply shaped Western intellectual history over the past hundred years.

### The subordination of space in social theory: 1880–1920

A distinctively different culture and consciousness of space, time, and modernity emerged in the decades preceding and following the end of the nineteenth century. Although actually lived experience may not have induced such a logical prioritization, at each level of philosophical and theoretical discourse, from ontology and epistemology to the explanation of empirical events and the interpretation of specific social practices, the historical imagination seemed resolutely to be erasing a sensitivity to the critical salience of human geographies. By the end of this period, the ascendance of a historicism of theoretical consciousness had reached such heights that the possibility of a critical human geography was made to appear inconceivable if not absurdly anachronistic to generations of Western Marxists and critical social scientists.

In the wake of these developments, the discipline of Modern Geography, which also took shape in the late nineteenth century, was squeezed out of the competitive battleground of theory construction. A few residual voices were heard, but the once much more central role of geographical analysis and explanation was reduced to little more than describing the stage-setting where the real social actors were deeply involved in making history. Social theorization thus came to be dominated by a narrowed and streamlined historical materialism, stripped of its more geographically sensitive variants (such as the utopian and anarchist socialisms of Fourier, Proudhon, Kropotkin, and Bakunin, as well as the pragmatic territorialism of social democracy); and a set of compartmentalized social sciences, each on its own becoming increasingly positivist, instrumental (in the sense of serving to improve capitalism rather than transform it), and, with a few exceptions, less attentive to the formative spatiality of social life as a template of critical insight.

In at least one of the disciplinary orthodoxies that consolidated around the turn of the century, the neo-classical economics of Marshall, Pigou, and others, the subordination of space was so great that its most influential theoreticians proudly produced visions of a depoliticized economy that existed as if it were packed solidly on to the head of a pin, in a fantasy world with virtually no spatial dimensions. Real history was also made to stand still in neo–classical economics and other variants of positivist and functionalist social science (as well as in some versions of modern Marxism), but the logic of time in the abstract was attended to through notions of causal process and sequential change, a comparative statics that rooted itself in natural science models of antecedent cause/ subsequent effect and the search for disengagingly independent variables. One might describe this as a mechanistic temporalism rather than an historicism of theory construction, but it tended just the same to expel spatiality.

Outside these disciplinary orthodoxies, in the two major streams of critical social theory, the imprint of historicism centered interpretation around the temporal dynamics of modernization and modernism. Modernization was conceptualized in Marxian political economy first of all in the revolutionary transition from feudalism to capitalism, the most epochal of all societal restructurings of the past and the defining moment for the conjunction of modernity with a particular mode of production. Marx built his critical understanding of capitalism from this transition and from the profound restructuring that took place during the age of revolution. But there was another problematic transition unfolding in the *fin de siècle* that required more than Marx's *Capital* to be understood theoretically and politically. The interpretation of this modernization process, its perils as well as its new possibilities, came to be dominated by a Leninist vanguard responding to the protracted rise of monopoly capital, corporate power, and the imperialist state.

There was great sensitivity to geographical issues in the writings of Lenin, Luxemburg, Bukharin, Trotsky, and Bauer, the key figures leading the early twentieth-century modernization of Marxism. Although not always in agreement, their collective works supplied a rich foundation for a Marxist theory of geographically (as well as historically) uneven development, one which built upon and extended in scope and scale that most geographically revealing of Marx's concepts, the syncretizing and synchronic antagonism between city and countryside, the agglomerative centre and the dissipative periphery. Nonetheless, *fin de siècle* Marxism remained solidly encased in historicism. The motor behind uneven development was quintessentially historical: the making of history through the unfettering struggle of social classes. The geography of this process, when it was seen at all, was recognized either as an external constraint or as an almost incidental outcome. History was the emotive variable container; geography, as Marx put it earlier, was little more than an 'unnecessary complication'. Like capitalism itself, the modern critique of capitalism seemed to be propelled through an annihilation of space by time. These early theories of imperialism would be resurrected later to assist in the reassertion of space in Marxist social theory, but this was more an act of desperation than inspiration, tapping one of the few areas of modern Marxist thought where geography seemed to matter.

In the critical social sciences, modernization was conceptualized around an at least superficially similar historical rhythm and rationalization, initiated with the origins of capitalism and the Industrial Revolution and moving through another troublesome transition at the end of the nineteenth century. Recomposing some of the same sources inspiring Western Marxism, from Kant and Hegel to Marx himself, critical social science characteristically defined this latter transition as the conclusive and categorical passage from Tradition to Modernity, *Gemeinschaft* to *Gesellschaft*, mechanical to organic solidarity. For the major theoreticians, Modernity (with a capital M) had indeed arrived, for better and worse. Above all, it demanded to be understood as the dominant theoretical and political referant, both at home and abroad.

What Marxists saw as the rise of imperialism via the internationalization of finance capital, the critical social scientists began to interpret as the time-lagged diffusion of development (as capitalist modernity) to the undeveloped, traditional, not yet fully modernized parts of the world. Here too there was a primarily Eurocentric vision which attached modernization everywhere to the historical dynamics of European industrial capitalism, to what Foucault described as 'the menacing glaciation of the world'. But whereas Marxism's theorization of history maintained a singular critical focus, social science rendered its critical historicism in many different versions: the methodological individualism of Max Weber, the sociology of collective consciousness of Émile Durkheim, the neo-Kantian scepticism of Georg Simmel, the intentional phenomenology of Edmund Husserl. In all these approaches, there was some attention given to human geography and to the geographically uneven development of society, but this geography of modernity remained essentially an adjunct, a reflective mirror of societal modernization.

While the two streams of critical social theory battled over the appropriate interpretation of history, the core modern movements of Marxism–Leninism and positivist Social Scientism engaged more pragmatically in changing it. Each coalesced around a different response to the *fin de siècle* restructuring of capitalism, creating hegemonic programmes for social progress that would shape the political cultures of the world throughout the twentieth century. The Marxist modern movement based itself in a revolutionary socialist strategy of vanguard action and a controlled territoriality of class struggle, a strategy that would, in due course, be successfully reinforced by events in Russia. An equally instrumental and opportunistic social science dedicated itself to the possibilities of scientifically planned reform primarily under the aegis of the liberal capitalist state, a visible hand of social guidance that would also be almost immediately reinforced by the successful reforms of liberalism in the 'progressive era' and, more ambiguously, in the liberal socialism associated with the rise of European social democracies. Both of these contemporaneous modern movements were to be shaken by crisis and doubt within their separate spheres of influence during the Great Depression and the Second World War, but they would emerge recharged, restructured, and even more antagonistically hegemonic in the 1950s.

The key argument I wish to establish in this admittedly broad and sweeping depiction of modernization and modernism is not only that spatiality was subordinated in critical social theory but that the instrumentality of space was increasingly lost from view in political and practical discourse. During the extended *fin de siècle*, the politics and ideology embedded in the social construction of human geographies and the crucially important role the manipulation of these geographies played in the late nineteenth–century restructuring and early twentieth–century expansion of capitalism seemed to become either invisible or increasingly mystified, left, right and centre.

Hidden within the modernity that was taking shape was a profound 'spatial fix'. At every scale of life, from the global to the local, the spatial organization of society was being restructured to meet the urgent demands of capitalism in crisis – to open up new opportunities for super–profits, to find new ways to maintain social control, to stimulate increased production and consumption. This was not a sudden development, nor should it be viewed as conspiratorial, completely successful, or entirely unseen by those experiencing it. Many of the avant-garde movements of the *fin de siècle* – in poetry and painting, in the writing of novels and literary criticism, in architecture and what then represented progressive urban and regional planning – perceptively sensed the instrumentality of space and the disciplining effects of the changing geography of capitalism. But within the consolidating and codifying realms of social science and scientific socialism, a persistent historicism tended to obscure this insidious spatialization, leaving it almost entirely outside the purview of critical interrogation for the next fifty years.

Why this happened is not easy to answer. How it happened is only now being discovered and explored in any detail. Part of the story of the submergence of space in early twentieth–century social theory is probably related to the explicit theoretical rejection of environmental causality and all physical or external explanations of social processes and the formation of human consciousness. Society and history were being separated from nature and naively given environments to bestow upon them what might be termed a relative autonomy of the social from the spatial. Blocked from seeing the production of space as a social process rooted in the same problematic as the making of history, critical social theory tended to project human geography on to the physical background of society, thus allowing its powerful structuring effect to be thrown away with the dirty bathwater of a rejected environmental determinism.

Another part of the story has to do with the modernist political strategies of the time. Here geography was given another reductionist interpretation and dismissal, not as external environment but as cognitive intuition. Those seeking the demise of capitalism, for example, tended to see in spatial consciousness and identity – in localisms or regionalisms or nationalisms – a dangerous fetter on the rise of a united world proletariat, a false consciousness inherently antagonistic to the revolutionary subjectivity and objective historical project of the working class. Only one form of territorial consciousness was acceptable – loyalty to the socialist state as soon as it came into being, but even that was considered only a temporary strategic convenience. Those seeking reformist solutions to the problems of capitalism were also uncomfortable with localisms and regionalisms which might too impatiently threaten the expectantly benevolent power of the capitalist state and instrumental social science. And there was the added threat of territorial nationalisms breaking the bonds of empire and cutting off the flow of profits so vital to 'metropolitan' reform. Here too only one form of territorial allegiance was cultivated and expected (i.e. to the national state), but national patriotism and citizenship were usually couched more in a cultural than a geographical identity and ideology, another example of the inherently spatial defined as something else. That the state was itself a socially produced space actively engaged in the reproduction of a particular social spatialization was thus rarely seen – and remained conspicuously absent from critical socialist and capitalist theories of state formation and politics.

## The mid-century involution of Modern Geography

By the 1920s, the isolation of Modern Geography and geographers from the production of social theory was well advanced. For most of the next fifty years, geographical thinking turned inwards and seemed to erase even the memories of earlier engagements with the mainstreams of social theorization. Only the ghost of Immanuel Kant effectively remained alive from that distanced past and its privileged apparition was used to lead the academic discipline of geography into further isolation. After all, who better to wrap Modern Geography in a warm cocoon of intellectual legitimacy than the greatest philosopher in centuries who also professed to be a geographer?[5]

Geography settled into a position within the modern academic division of labour that distinguished it (and rationalized its distinction) from both the specialized and substantive disciplines of the natural and human sciences (where theory was presumed to originate) and from history, its allegedly co-equal partner in filling up 'the entire circumference of our perception', as Kant put it. Geography and history were ways of thinking, subjective schemata which co-ordinated and integrated all sensed phenomena. But by the 1920s, putting phenomena in a temporal sequence (Kant's *nacheinander*) had become much more significant and revealing to social theorists of every stripe than putting them beside each other in space (Kant's *nebeneinander*). History and historians had taken on a crucial interpretive role in modern social theory: an integrative and cross-disciplinary responsibility for the study of development and change, modernity and modernization, whether expressed in the biography of individuals, the explanation of particular (historical) events, or the tumultuous transformations of social systems. The historian as social critic and observer, history as a privileged interpretive perspective, became familiar and accepted in academic and popular circles. In contrast, geography and geographers were left with little more than the detailed description of outcomes, what came to be called by the chroniclers of the discipline the 'areal differentiation of phenomena' (Hartshorne, 1939, 1959).

The exceptional theoretical acquiescence of mid-century human geography was a slideway to disciplinary involution. Here and there, a few geographers individually contributed to theoretical debates in the social sciences and scientific socialism, drawing mainly on the continuing strengths of physical geography and the occasional appeals of historians to limited environmental explanations of historical events. But the discipline as a whole turned inwards, abstaining from the great theoretical debates as if a high wall had been raised around it.

With its Kantian *cogito* mummified in neo-Kantian historicism, Modern Geography was reduced primarily to the accumulation, classification, and theoretically innocent representation of factual material

---

5. Kant helped to make ends meet by lecturing on geography at the University of Konigsberg for nearly forty years. He gave his course forty-eight times, lecturing more often only on logic and metaphysics. Kant saw geography – mainly physical geography – as 'the propaedeutic for knowledge of the world' (See May, 1970, 5). May opens his interesting account with another Kantian assertion: 'The revival of the science of geography .... should create that unity of knowledge without which all learning remains only piece-work.'

describing the areal differentiation of the earth's surface – to the study of outcomes, the end products of dynamic processes best understood by others. Geography thus also treated space as the domain of the dead, the fixed, the undialectical, the immobile – a world of passivity and measurement rather than action and meaning. Accurate packages of such geographical information continued to be of use to the state, in the East and West, for military intelligence, economic planning, and imperial administration. These three arenas of intelligence, planning, and administration defined an 'applied' geography almost by default, cementing a special relationship with the state that probably arose first in an earlier age of imperial exploration. The majority of the most prominent mid-century geographers in the United States of America were tied in one way or another with intelligence-gathering activities, especially through the Office of Strategic Services, the progenitor of the CIA, and there still remains an office of 'The Geographer' in the State Department in recognition of dedicated and disciplined service. Without undue exaggeration, the French radical geographer Yves Lacoste – one of those who interviewed Foucault on geography – entitled his book on the field *La Géographie: ça sert, d'abord, à faire la guerre* (Lacoste, 1976).

Given this attachment to the state, it is not surprising that the subfield of political geography generated the most active attempts at theorization. Sir Halford Mackinder's notion of the Eurasian 'Heartland' as the 'geographical pivot of history' (1904) and his active participation in redrawing the map of Europe after the First World War (see Mackinder, 1919) established and legitimized geopolitics as the primary practical and theoretical focus of human geography. This centrality would last through the interwar years, at least until the aberrant episode of German *geopolitik* made non-fascist geographers think twice about venturing too far into the realms of political theory building. With its theoretical fingers burnt again, human geography as a whole retreated into the calmer climes of mere description, while political geography became what some called the discipline's moribund backwater.

Backed into its neo-Kantian cocoon, the explanation of human geographies took several different forms. One emphasized the old environmental 'man/land' tradition and sought associations between physical and human geographies on the visible landscape, either via the influences of the environment on behaviour and culture or through 'man's role in changing the face of the earth' (Thomas, 1956). Another concentrated on the locational patterns of phenomena topically organized to reflect the established compartments of modern social science. This defined specialized fields of economic, political, social, cultural, and much later, behavioural or psychological geography, but not, one might add, a geography based in political economy. A third approach aimed at synthesizing everything in sight through a comprehensive and typically encyclopedic regionalization of phenomena, an approach considered by most mid-century geographers to be the distinctive essence of the discipline. Finally, a historical geography roamed freely through all three of these approaches tracing the human geographies of the past as a temporal sequencing of areal differentiation and basking in the intellectual legitimacy and power of the historical imagination. Characterizing every one of these forms of geographical analysis, from the most tritely empirical to the most insightfully historical, was the explanation of geographies by geographies, geographical analysis turned into itself, the description of associated outcomes deriving from processes whose deeper theorization was left to others.

While this involution occurred, the main currents of Western Marxism and critical social science lost touch with the geographical imagination. There were a few small pockets of provocative geographical analysis and theorization that survived through this mid-century spatial acquiescence: in the evolutionary urban ecology of the Chicago School; in the urban and regional planning doctrines that were consolidated in the interwar years (see Friedmann and Weaver, 1979; Weaver, 1984); in the regional historiography and attention to environmental detail of the French *Annales* School, with its continuation of the traditions of Vidal de la Blache; among certain North American and British historians still inspired by Frederic Jackson Turner and other frontier theorists or by the Marxist theories of imperialism; in the work of Antonio Gramsci on the regional question, local social movements, and the capitalist state. For the most part, however, what was being kept alive in these residual pockets was the geographical imagination in retreat of the extended *fin de siècle*. Relatively little that was new was added after the early 1930s and even the preserved remains of the past were encased in an ascendant and confining historicism which consigned its geography to the background of critical social discourse.

In any case, nearly all these residual pockets were to dwindle in impact and importance through the Great Depression and the war years so that by 1960 their specifically geographical insights were only dimly perceptible. At this point, with postwar recovery and economic expansion in full flow throughout the advanced capitalist and socialist world, the despatialization of social theory seemed to be at its peak. The geographical imagination had been critically silenced. The discipline of Modern Geography was theoretically asleep.

*Uncovering Western Marxism's spatial turn*

Very little has yet been written on the despatialization of social theory up to the 1960s. The sounds of silence are difficult to pick up. The works of Perry Anderson (1976, 1980, 1983), however, offer an excellent critical survey of Western Marxism that almost inadvertently chronicles the loss of spatial consciousness in mainstream Marxism after the Russian Revolution and simultaneously prepares the way, again without necessarily intending to do so, for understanding how and why the pertinent spatiality of social life began to be rediscovered in the late 1960s. To end this chapter and to introduce the next, I will use Anderson's work to locate the origins of what would eventually become a lively and productive encounter between Western Marxism and Modern Geography.

From 1918 to 1968, Anderson argues, a new 'post-classical' Marxist theory crystallized to redirect historical materialist interpretations of what I have called modernity, modernization and modernism. This retheorization was geographically unevenly developed, finding its primary homelands in France, Italy, and Germany, 'societies where the labour movement was strong enough to pose a genuine revolutionary threat to capital' (1983, 15). In Britain and the United States no such revolutionary challenge was apparent, while in the east a rigid Stalinist economism left little room for redirection and reinterpretation. For Anderson, the founding fathers of this countercurrent were Lukacs, Korsch, and Gramsci, while following in their wake were the more modern figures of Sartre and Althusser in France; Adorno, Benjamin, Marcuse and others associated with the Frankfurt School in Germany; and Della Volpe and Colletti in Italy.

This Latinate and Frankfurtized movement shifted the institutional and intellectual terrain of Marxist theory, rooting it more than ever before in university departments and research centres, and in a resurgent interest in philosophical discourse, questions of method, a critique of bourgeois culture, and such subjects as art, aesthetics, and ideology (which were lodged in the classically neglected realms of capitalism's superstructure). More traditional infrastructural themes having to do with the inner workings of the labour process, struggles at the workplace over the social relations of production, and the 'laws of motion' of capitalist development tended to be given relatively less attention. The same was true for more conventionally political (and I might add, geographical) topics as the organization of the world economy, the structure of the capitalist state, and the meaning and function of national identity, although here even the classical theorists were often neglectful as well.

To Anderson, Marxism seemed to be moving backwards, from economics through politics to focus on philosophy, reversing the consummate path taken by Marx. Philosophically understanding the world had taken precedence over changing it. But by the 1970s, this 'grand Western Marxist tradition' had 'run its course' and was being replaced by 'another kind of Marxist culture, primarily oriented towards just those questions of an economic, social or political order that had been lacking from its predecessors' (1983, 20). This restructured Western Marxism also took on a different geography, becoming centred in the English-speaking world rather than in Germanic or Latin Europe. As a result, 'the traditionally most backward zones of the capitalist world, in Marxist culture, have suddenly become in many ways the most advanced' (Ibid., 24).

Anderson's *In the Tracks of Historical Materialism* (1983) reflects upon his earlier works, his successful and not so successful projections of the fate of this restructured Marxism, with its new zest for the concrete and its re-centred geography.[6] Following an introductory personal scorecard on 'Prediction and Performance', Anderson first retraces and expands upon the essentially French debates on 'Structure and Subject' that he argues led to the current crisis of Latin Marxism; and then moves through a review of German Marxism (focusing mainly on Habermas) to elucidate the even more perplexing debates on the changing relations between 'Nature and History'. Hidden within these chapters, however, is a lateral story-line that Anderson edges towards but fails to see, an emerging postmodern discourse that was seeking not to dismiss Marxism as a critical theory but to open it up to a necessary and overdue spatialization, to a materialist interpretation of spatiality that would match its magisterial historical materialism.

This initial assertion of a postmodern critical human geography was almost entirely confined to the French Marxist tradition, which had always been more open to the spatial imagination than its Anglo-American and German counterparts. Sartre's 'search for a method' in his increasingly Marxist existentialism and Althusser's anti-historicist re-reading of Marx were the primary pre-texts for this Gallic spatialization. The Marxified phenomenological ontology of Sartre represented an hermeneutic that centred on the subjectivity, intentionality, and consciousness of knowledgeable human agents engaged not only in making history but also in shaping the political culture of everyday life in

---

6. It is not geographically inconsequential that *In the Tracks...* is a publication of the Wellek Library Lectures, given by Anderson in 1982 at the University of California, Irvine. It is also interesting to note how filled these lectures were with resplendent spatial metaphors, despite their underlying historicism. Anderson describes his work as a 'cadastral survey' of the shifting 'terrain', an exploration of the 'shaping of a new intellectual landscape', a 'changing map of Marxism'.

modern capitalist society. Althusser's structuralism, in contrast, emphasized the more objective conditions and social forces which shape the underlying logic of capitalist development and modernization. Each contributed to channelling post-war French Marxism into two discordant streams, split by opposing views of the structure–subject relation but both peculiarly open to the possibility of spatialization.

The crisis of French Marxism that Anderson sadly describes as a 'massacre of ancestors' was a crisis of disillusionment that 'exploded' French Marxism into a multitude of fragments, obliterating the orthodoxies of the immediate and more distant past.[7] Faced with this unprecedented heterogeneity, fragments flying every which way (including antagonistic departures from Marxism entirely), Anderson mourned the symptomatic loss of faith. Sartre would turn in his last years to a 'radical neoanarchism', Althusser and Poulantzas to exasperated lamentations on the absence of a theory of politics and the state in historical materialism. The bedevilling (and also 'neo-anarchistic') Foucault – with Derrida and many others – would dilute and diminish Marxism still further, in Anderson's view, promoting a contagious 'randomization of history' and celebrating the triumphant ascendancy of a poststructuralist (and by implication post–Marxist) episteme. With splendid irony, however, Anderson finds one telling exception to this 'precipitous decline' in French Marxism:

> No intellectual change is ever universal. At least one exception, of signal honour, stands out against the general shift of positions in these years. The oldest living survivor of the Western Marxist tradition I discussed, Henri Lefebvre, neither bent nor turned in his eighth decade, continuing to produce imperturbable and original work on subjects typically ignored by much of the Left. The price of such constancy, however, was relative isolation. (1983, 20)

Lefebvre is discovered seemingly out of nowhere. Anderson gives little attention to him in his earlier works and little more is mentioned of Lefebvre in the discussion of the contemporary decline of French Marxism. What was of such signal honour, exceptional constancy, imperturbable originality in the works of Lefebvre? I suggest that this perhaps least known and most misunderstood of the great figures in twentieth–century Marxism has been, above all else and others, the incunabulum of postmodern critical human geography, the primary source for the assault against historicism and the reassertion of space in critical social theory. His constancy led the way for a host of other attempted spatializations,

---

7. See Lefebvre's discussion of *Le Marxisme éclaté* in *Une pensée devenue monde ...*
*faut-il abandonner Marx?* (Lefebvre, 1980), pp. 16–19 especially.

from Sartre, Althusser, and Foucault to Poulantzas (1978), Giddens (1979, 1981, 1984), Harvey (1973, 1985a, 1985b), and Jameson (1984). And he remains today the original and foremost historical and geographical materialist.

Anderson misses this creative recomposition of Western Marxism taking place amidst the great French deconstruction which followed the explosion at Nanterre in 1968. His interpretations of such key figures as Sartre, Althusser, and Foucault too quickly dismiss as retreats from politics their creative, but not entirely successful, ontological struggles with the spatiality of existential being, modernity, and power; and he does not see at all the anglophonic Marxist Geography that Lefebvre and other French Marxists helped to stimulate. Despite his sensitive tapping of francophone Marxist traditions, Anderson still seems trapped in the 'historically centred Marxist culture' of the anglophone world, in which he claims that 'theory is now history, with a seriousness and severity it never was in the past; as history is equally theory, in all its exigency, in a way it typically evaded before.' Adding to history 'and geography' would have made a world of difference.

# 2

# Spatializations: Marxist Geography and Critical Social Theory

The dialectic is back on the agenda. But it is no longer Marx's dialectic, just as Marx's was no longer Hegel's.... The dialectic today no longer clings to historicity and historical time, or to a temporal mechanism such as 'thesis–antithesis–synthesis' or 'affirmation–negation–negation of the negation' ... To recognise space, to recognise what 'takes place' there and what it is used for, is to resume the dialectic; analysis will reveal the contradictions of space. (Lefebvre, 1976, 14 and 17)

The discourses of Modern Geography and Western Marxism rarely crossed paths after their formative period in the extended *fin de siècle*. Geography isolated itself in a tight little island of its own, building a storage–house of factual knowledge that was only occasionally broadcast into the public domain. Marxism meanwhile stashed away the geographical imagination in some superstructural attic to gather the dust of discarded and somewhat tainted memories. Only in France, as we have seen, did a vibrant spatial discourse survive the mid-century despatialization, keeping alive a debate that seemed to have disappeared entirely in other, non-Latinate Western Marxisms.

In the early 1970s, however, a resolutely Marxist geography began to take shape from a sudden infusion of Western Marxist theory and method into the introverted intellectual ghetto of anglophonic Modern Geography. It formed a vital part of a nascent critical human geography which arose in response to the increasingly presumptive and theoretically reductionist positivism of mainstream geographical analysis (Gregory, 1978). Although this newborn Marxist geography tended to be inward-looking, unsettled in its critical stance, and perhaps therefore largely unnoticed outside the disciplinary discourse, it shook the foundations of Modern Geography and initiated a debate that would eventually extend well beyond the disciplinary cocoon.

Through the decade of the 1970s, Marxist geography remained peripheral to Western Marxism, built almost entirely on a one-way flow of ideas, an increasing Marxification of geographical analysis and explanation. After 1980, however, the scope of the encounter between Modern Geography and Western Marxism changed as the flow of ideas and influences began to move, ever so slightly, in both directions. As we move closer to the next *fin de siècle*, a broader and deeper critical debate on the appropriate theorization of the spatiality of social life is reaching into and challenging long-established traditions in Western Marxism, while simultaneously forcing a major rethinking of the conceptual and institutional frameworks of Modern Geography as well.

One of the most explicit and focused expressions of this widening critical debate has been the assertion of a profoundly spatialized historical materialism. As David Harvey, a pre-eminent figure in the development of Marxist geography from the start, has recently argued: 'The historical geography of capitalism has to be the object of our theorizing, historico-geographical materialism the method of inquiry' (1985, 144). This historico-geographical materialism is much more than a tracing of empirical outcomes over space or the description of spatial constraints and limitations on social action over time. It is a compelling call for a radical reformulation of critical social theory as a whole, of Western Marxism in particular, and of the many different ways we look at, conceptualize, and interpret not only space itself but the whole range of fundamental relationships between space, time, and social being at every level of abstraction. As Lefebvre suggests in the quotation at the head of this chapter, it is an invitation to 'resume the dialectic' on a different interpretive terrain.

But something else has been happening since 1980 alongside the initial calls for an historical and geographical materialism from within Marxist geography. There has been, first of all, an unprecedented generalization of the debate on the theorization of space and time, geography and history, not only in social theory but in broader realms of critical discourse in art, architecture, literature, film, popular culture, and contemporary politics. Today, the debate has expanded well beyond the confines of Marxist geography and drawn into the discussion a range of critical participants who no longer fit comfortably within the conventional labels of either 'geographer' or 'Marxist'.

At the same time, it is becoming increasingly clear that the insertion of space into historical materialism and into the wider frameworks of critical theory is not just a matter of simple incremental adaptation, the incorporation of still another new variable or model into the old and unquestioned master narratives. Critical theory and Western Marxism have been so muted with regard to spatiality for so long that the inclusion of a theoretically meaningful spatial dimension may not be possible without shattering many well-established interpretive assumptions and approaches, especially those associated with the deeply engrained primacy of historical versus geographical modes of explanation and critique. Similarly, Modern Geography has been so introverted and cocooned with respect to the construction of critical social theory and so confined in its definition of historical geography that it may also be incapable of adjusting to the contemporary reassertion of space without a radical deconstruction and reconstitution.

In effect, the long delayed encounter between Modern Geography and Western Marxism is now threatening to become mutually transformative. It thus becomes particularly important to retrace the origins and development of Marxist geography and to move along the rediscovered tracks that have led to the call for an historico-geographical materialism and to the postmodern explosion of the critical debate on the theorization of space.

## Finding Roots: Spatiality in the French Marxist Tradition

Marxist geography developed primarily in anglophonic countries and out of the turbulence of the 1960s, when virtually every social science discipline seemed to spawn a reawakened radical fringe. The formative theorizations of Marxist geography, however, were predominantly francophonic and reflected the peculiar centrality which space had reassumed in the French intellectual traditions of the twentieth century.[1] A partial explanation for this peculiar centrality, especially on the left, can be found in the history of French Marxism.

Marxism in France was relatively late in developing when compared to Britain, Germany, and the United States. This has been attributed largely to the powerful inheritance of earlier French socialist thought, which continued to offer an attractive and indigenous political alternative

---

1. I use 'reassume' to cover what some observers consider to be a 'drying up' of this French emphasis on spatiality during the nineteenth century. Gregory (1978, 38) writes: 'In France the spatial dimension had been an insistent element of studies in political economy since the end of the seventeenth century, and this lasted well into the eighteenth. By the early 1800s, however, it was beginning to falter', in part, Gregory argues, due to the rise of a Ricardian political economy and, later, to the conquest of Comtean positivism. This produced a 'lacuna' between the early growth of a spatialized political economy and the twentieth–century development of the epistemology to which it would eventually subscribe. Lacuna or not, the underlying point is the extraordinary depth and continuity of the French spatial tradition when compared with German idealist philosophies and British political economy.

to the left well into the twentieth century (Kelly, 1982; Poster, 1975).
When French Marxism did expand under the immiserating conditions of global economic crisis in the 1920s and 1930s, it was shaped by
several distinctively local circumstances. For example, it built upon a
heritage of political and social theory which, from Saint Simon, Fourier
and Proudhon, to the anarchist geographers Kropotkin and Reclus,
contained a sensitive and persistent emphasis on the politics of spatiality
and territorially-based communalism (Weaver, 1984). Although not
phrased in so many words, the political strategies of these utopian and
libertarian socialists revolved around the need to recapture social control
over the production of space from an expansive capitalism and an
equally expansive and instrumentalist capitalist state.

French Marxism also grew with less of the anti-spatial bias that had
already become engrained in the more 'advanced' national Marxisms of
the other Western industrial countries. Much of this anti-spatialism originated in Marx's 'double inversion' of Hegel, a critical turnaround that I
will refer to again in later chapters. In grounding the Hegelian dialectic
in material life, Marx not only responded to Hegelian idealism, denying
the spiritual navigation and determination of history, he also rejected its
particularized spatial form, the territorially defined state, as history's
principal spiritual vehicle.

Standing the Hegelian dialectic 'on its feet' was thus both a denial of
idealism and a specific rejection of territorial or spatial fetishism, a
hermeneutic in which history was determined by an innately given
spatial consciousness whether focused on the state, cultural nationalism,
regionalism, or local communalism. In the Marxian dialectic, revolutionary time was re-established, with its driving force grounded in class
consciousness and class struggle stripped of all spatial mystifications.
The Marxian inversions were used to squeeze out Hegelian influences
around the turn of the century, even in Germany, and implanted a
forbidding theoretical and political anti-spatialism. The early expansion
of Marxism in France, however, coincided with a major Hegelian
revival, a reinvestiture that carried with it a less expurgated sensitivity to
the spatiality of social life.

There seems to be a convincing basis for claiming that French Marxism, from the onset, was more inclined toward an explicit spatial
perspective and theorization than Marxisms elsewhere.[2] Throughout the

---

2. Italian Marxism was probably similarly inclined. There may be some basis in arguing
for a broader Latin tradition of spatiality, but I am not prepared to make this argument
here, other than to note the important role of Antonio Gramsci in the development of a
historical-geographical materialism. Gramsci is so much more spatial than the other founding fathers of Western Marxism, Karl Korsch, Ernst Bloch, and Georg Lukacs. I will return
to Gramsci's geography in the next essay.

twentieth century, French critical thought, whatever its primary source, preserved an ongoing spatial discourse – from Cubism and the Surrealist movement, through the bifurcated currents of Althusserian structuralism and Sartrean existentialism, to the contemporary poststructuralist and postmodernist debates. But this lively geographical imagination rarely travelled well across the language barrier, even among those drawing most heavily on the French critical tradition.

It is no surprise, then, that the distinctively French debates on the theorization of space rarely penetrated the more historicist armour of other, non-Latin Marxisms, and are frequently omitted entirely in Anglo-American historical writings on French Marxism. This hidden history of spatialization is most acutely illustrated and brought back into view by focusing on the career of Henri Lefebvre, whose life has now spanned from one *fin de siècle* (his date of birth is usually given as 1901) to another. Lefebvre was perhaps the most influential figure shaping the course and character of French Marxist theory and philosophy from the early 1930s to at least the late 1950s. He became, after the 1950s, the leading spatial theoretician in Western Marxism and the most forceful advocate for the reassertion of space in critical social theory. Yet only in the present decade have his remarkable accomplishments begun to be fully recognized and appreciated in the historically centered Marxist culture of the anglophone world.

Hegelian influences figure prominently in Lefebvre's early Marxism. With his associate, Norbert Guterman, Lefebvre published the first French translation of key sections of Marx's 1844 economic and philosophical manuscripts and, in a series of anthologies accompanied by extensive editorial notes and discussion, introduced to a French audience many other works by Marx and Engels as well as Lenin's *Philosophical Notebooks*. The latter was Lenin's critical appreciation of Hegel, and Lefebvre's translation contributed significantly to the Hegelian revival.[3]

In his own elaboration of the Hegel–Marx relation, Lefebvre sought to retain a strand of 'objective idealism' within the materialist dialectic, to encourage attention to contradictions in thought and consciousness as well as to the material bases of contradictions in concrete reality and history. *La Conscience mystifiée*, the title of one of his earliest original works, set up an insistent theme for Lefebvre and for many other French

---

3. As noted in Jay (1984, 293), Lefebvre's first contact with Hegel came from his early involvement in the Surrealist movement and with one of its leading figures, André Breton. In his autobiographical *Le Temps des méprises* (1975, 49), Lefebvre describes an encounter with Breton in 1924: 'He showed me a book on his table, Vera's translation of Hegel's *Logic*, a very bad translation, and said something disdainfully of the sort: "You haven't even read this?" A few days later, I began to read Hegel, who led me to Marx.'

Marxist writers.[4] Lefebvre explicitly accepted Marx's argument about the primacy of material life in the production of conscious thought and action – that social being produced consciousness rather than the reverse – but refused to reduce thought and consciousness to a determined aftergloss or mechanical ideation. These ideas grew out of Lefebvre's attachment to the French Surrealist movement and his early existentialism. They also may have come as close to provoking a French counterpart to the contemporaneous Frankfurt School as anything else in France during the interwar period.

In a pattern which would characterize his work for over fifty years, Lefebvre took a stance against dogmatic reductionism in the interpretation of Marx. He argued instead for a flexible, open, and cautiously eclectic Marxism able to grow and adapt without predetermined truncation. This anti-reductionism made him the most influential early critic of Stalinist economism through his widely circulated and translated text, *Le Matérialisme dialectique*, originally published in 1939 (English trans. 1968), one of the most widely read introductions to Marxism ever written. Later, both existentialism and structuralism (1971) would receive Lefebvre's sting through biting critiques of the early Sartre and the ascendant Althusser, aimed again at the dangers of reductive 'totalizations'.

In seeking dialectically to combine the relational contradictions of thought and being, consciousness and material life, superstructure and economic base, objectivity and subjectivity, Lefebvre was not alone. But he was the first to apply this reformulated dialectical logic to combine the strengths and shed the weaknesses of existential phenomenology and Althusserian structuralism, seeing in these two major twentieth-century philosophical movements a creative opportunity to improve and strengthen Marxism (while at the same time rejecting their restrictive theoretical rigidities). Over the past thirty years, Lefebvre has drawn selectively from these movements in an insistent attempt to recontextualize Marxism in theory and praxis; and it is within this recontextualization that we can discover many of the immediate sources of a materialist interpretation of spatiality and hence of the development of Marxist geography and historical-geographical materialism.

Lefebvre's theorization of space cannot be summarized easily, for it is

---

4. Lefebvre and Guterman (1936). The main argument of this work was that all forms of consciousness, individual and collective, are manipulated under capitalism to hide the fundamental mechanisms of surplus-value extraction and accumulaticn. This meant that the working class itself was also likely to be unaware of the means of its own exploitation, trapped in an equally mystified consciousness at least until it discovered how to lift the veil of this instrumental mystification. Publication of the book caused widespread political controversy within the Communist Party, especially since it was written by the Party's leading theoretician at the time.

embedded in an extraordinary number of published works which touch upon virtually every aspect of social theory and philosophy. When probed for the origins of his interest in spatiality, Lefebvre points to the influence of his Occitanian birthplace and his frequent returns home to observe the massive changes taking place in rural land and life under the impress of state-directed spatial planning.[5] His more explicit theorizations, however, evolve through a series of what he called 'approximations' of a central thesis, a rather slippery path that seemed to confuse many of his followers and critics alike.

The first approximation took the form of a *Critique de la vie quotidienne*, everyday life in the modern world (1946b, 1961, 1968a), an extension of his arguments in *La Conscience mystifiée* (and a prefiguration of Braudel's work on the structures of everyday life and *mentalité*). Lefebvre concentrated attention on the characteristic features of the modernized capitalism that consolidated around the turn of the century in what he called a 'bureaucratic society of controlled consumption' choreographed by the capitalist state – in essence, an instrumentalized 'spatial planning' which increasingly penetrated into the recursive practices of daily life. When Lefebvre is recognized at all by contemporary Western Marxists, he is remembered most for this innovative first approximation.

By the time his projected trilogy on everyday life was completed, Lefebvre had already begun to recast his work around such themes as the struggle for control over 'the right to the city' (*Le Droit à la ville*, 1968b); the urbanization of consciousness; and the degree to which the transformation of capitalism accordingly required an 'urban revolution' (1970a). Interwoven with these urban approximations was an exploration of the 'repetitive versus the differential', an exploration of the homogenizing effects of capitalism, its capacity to obliterate differences

---

5. See Lefebvre (1975, 9) for his response to the question of how he developed his interests in space. Particularly influential was the construction of the new town of Lacq-Mourenx in the Pyrenées-Atlantiques, and what he called 'the emergence of a new social and political practice' associated with the establishment of DATAR (Délégation à l'Aménagement du Territoire et à l'Action Régionale). A recent translation of an interview with Lefebvre (Burgel, et al., 1987) presented these responses to the same question: 'The point of departure for me was the work of DATAR.... Something new was happening; an idea of spatial planning and practice was born.... After all, the creation of new towns and the redevelopment of existing ones was quite a new approach compared with classic descriptions of urban phenomena. However, my initiation was neither from the point of view of philosophy, nor sociology, though these were present implicitly, nor was it historical or geographical. Rather it was the emergence of a new social and political practice. DATAR aimed to reorganise France from questionable, sometimes catastrophic, perspectives.... It was certainly an original French phenomenon. I don't know of many other countries that have gone beyond the stage of financial planning in their budgets to actually planning their space.' (28)

or what more contemporary critical scholars might call *différance*(1970b).

As Lefebvre kept trying to explain, his use of the terms 'urban' and 'urbanism' stretched well beyond the immediate confines of cities. Urbanization was a summative metaphor for the *spatialization* of modernity and the strategic 'planning' of everyday life that has allowed capitalism to survive, to reproduce successfully its essential relations of production. After looking back to Marx's writings on the city (1972), Lefebvre began to articulate more clearly his central thesis on spatiality and social reproduction in *La Survie du capitalisme* (1973; English translation, 1976a) and his masterwork, *La Production de l'espace* (1974). The very survival of capitalism, Lefebvre argued, was built upon the creation of an increasingly embracing, instrumental, and socially mystified spatiality, hidden from critical view under thick veils of illusion and ideology. What distinguished capitalism's gratuitous spatial veil from the spatialities of other modes of production was its peculiar production and reproduction of geographically uneven development via simultaneous tendencies toward homogenization, fragmentation, and hierarchization – an argument that resembled in many ways Foucault's discourse on heterotopias and the instrumental association of space, knowledge, and power. 'This dialectised, conflictive space is where the reproduction of the relations of production is achieved. It is this space that produces reproduction, by introducing into it its multiple contradictions' (Lefebvre, 1976a, 19) – contradictions which must be analytically and dialectically 'revealed' to enable us to see what is hidden behind the spatial veil.

This was the most forceful theoretical and political assertion ever made in Western Marxism of the importance of spatiality and the existence of an intrinsic spatial problematic in the history of capitalism. And it was destined to be either ignored or misunderstood by most Marxists, for it represented a shattering attack against the orthodoxies of Marxist thought and above all to its hegemonic historicism. Kelly (1982) and Hirsch (1981) do not even mention Lefebvre's writings on urbanism, much less see his spatial turn in their recent histories of French Marxism, while Poster (1975) only briefly and with puzzlement discusses Lefebvre's 'new praxis' of urbanism in conjunction with the events of May 1968 – which began in Nanterre, where Lefebvre was teaching.[6]

---

6. After the end of the Paris uprisings in 1968, which many saw as a political test of Lefebvre's ideas, even some of his closest followers began to abandon him to leap onto what were then the two biggest bandwagons of French Marxism, Althusserian structuralism and Sartrean existentialism. Significantly, it was this post-1968 Lefebvre-you-have-failed-us mood which coloured the translation of Lefebvre's work into anglophonic Marxism (for example, Castells, 1977).

Even today, the most extensive treatment of Lefebvre's work written in English (Saunders, 1986) fails to list the pivotal *La Production de l'espace* and his more recent writings on the state and the world economy (Lefebvre, 1976–78, 1980) in its bibliography. Even worse, Saunders uses his miscomprehension of Lefebvre's 'spontaneity' and 'speculative' style to argue for the abandonment of a spatial emphasis in anglophonic urban sociology.

Nonetheless, from the early 1970s to the present, the encounter between Modern Geography and Western Marxism, and the formation and reformation of Marxist geography, develops around and towards the reconfigured dialectic which Lefebvre describes in the passage that introduces this chapter. It is an increasingly spatialized dialectic, an insistent demand for a fundamental change in the ways we think about space, time, and being; about geography, history, and society; about the production of space, the making of history, and the constitution of social relations and practical consciousness. The Lefebvrian 'assertion' is the key moment in the development of an historico-geographical materialism and will be returned to again and again in subsequent essays.

### Adding Marx to Modern Geography: the First Critique

The anglophonic contribution to Marxist geography primarily hinged upon the reconnection of spatial form to social process, an attempt to explain the empirical outcomes of geographically uneven development (what geographers innocently called 'areal differentiation') through its generative sources in the organizational structures, practices, and relations that constitute social life. This reconnection was asserted in principle during the late 1950s, when the so-called 'quantitative-theoretical revolution' arose from within Modern Geography's introverted and virtually atheoretical cocoon. This increasingly technical and mathematized version of geographical description, however, differed only superficially from the neo-Kantian tradition that helped to justify the isolation of geography from history, the social sciences, and Western Marxism. It grounded explanation primarily in social physics, statistical ecologies, and narrow appeals to the ubiquitous friction of distance. But after all was said and done, outcomes continued to explain outcomes in an infinite regression of geographies upon geographies, one set of mappable variables 'explaining' another through the 'goodness' of fit. The adopted positivist stance, even when humanized somewhat through 'behavioural' approaches and phenomenological fine-tuning, merely re-legitimated Modern Geography's fixation on empirical appearances and involuted description.

But a crack had been opened as anglophonic geographers collectively became more aware of their isolation and began to search for new attachments outside their old cocoon. There were 'space invaders' everywhere in the 1960s, especially in North America, as more theoretically minded geographers wandered into every disciplinary location they could find, from mathematical topology and analytical philosophy to neo–classical economics and cognitive psychology. The world outside, however, was changing rapidly. Cities, regions, and states were increasingly riddled by conflict, crisis, and the beginnings of a profound restructuring. The academic environment was becoming more politicized – and acceptably 'radical' – than it had been for many years in North America and Great Britain; and theoretical discourse had begun to turn significantly against positivism and toward critical alternatives drawn from the 'grand houses' of continental European social theory.

In this context of change, parts of Modern Geography also began to radicalize, led by contributors to the new journal *Antipode* and inspired by a series of leftward turns taken by some of the most prominent anglophonic geographers of the time. David Harvey's dramatic shift in direction, from the positivist ecumenicism of *Explanation in Geography* (1969) to the avowedly Marxist *Social Justice and the City* (1973), was particularly pathbending and influential, especially to the generation of young geographers so recently taught by their professors to pay close attention to Harvey's work. Modern Geography would never be the same after Harvey's provocative redirection.

Although more heterogeneous at first, radical geography moved quickly toward a dedicated Marxification of geographical analysis, again spearheaded by Harvey. Historical materialism became the preferred route to connect spatial form with social process, and thereby to combine human geography with class analysis, the description of geographical outcomes with the explanations provided by a Marxian political economy. One by one, the familiar themes of Modern Geography were subjected to a Marxist analysis and interpretation: the patterns of land rent and land use, the variegated forms of the built environment, the location of industry and transport routes, the evolution of urban form and the ecology of urbanization, the functional hierarchy of settlements, the mosaic of uneven regional development, the diffusion of innovations, the evocations of cognitive or 'mental' maps, the inequalities in the wealth of nations, the formation and transformation of geographical landscapes from the local to the global.

This new approach to geographical explanation had several important features. At its core was a radical political economy based essentially on Marx's *Capital,* with occasional derivations from *Grundrisse* and later theories of imperialism. But carried along with these conventional

sources were three more contemporary variations, often confusingly
intermingled: 1) a largely British Marxist tradition, perhaps more rigidly
historicist than any other, averse to speculative theorizing and deeply
attached to pragmatic empirical analysis; 2) a brash and venturesome,
perhaps characteristically New World-based, 'neo-Marxism' springing
from a presumed necessity to update Marxian principles and drawing
from more unconventional sources of insight; and 3) an ebbing but still
influential French Marxist tradition, split into several streams (structuralist, existentialist, and various iterations of each), virtually incomprehensible to the unbending historicists but enticingly inspirational to
neo-Marxism.

The structuralist 'reading' was particularly attractive to Marxist
geography, for it provided an apparently rigorous epistemological
rationalization for digging under the surface appearance of phenomena
(spatial outcomes) to discover explanatory roots in the structured and
structuring social relations of production. This fitted perfectly into the
formative project and logic of Marxifying geographical analysis. Structuralism, primarily of the Althusserian variety, was also instructively
anti-positivist; it supplied an effective foil against the theoretical human-
ism that was inspiring alternative behavioural and phenomenological
critiques of positivist geography, critiques which often adopted a decidedly
anti-Marxist stance; and it opened up the superstructural realm which
seemed to contain so much of what geographers looked at. Add to all
this structuralism's programmatic attack on historicism and its enthusiasm for compelling spatial metaphors, and it becomes easier to understand its powerful attraction to Marxist geographers. An alluring
pathway was provided for geography to enter into the mainstream of
critical theoretical debates in Western Marxism instead of forever waiting in the surrounding undergrowth. The other possible paths which
appeared – via existential phenomenology or the Frankfurt School tradition of critical theory – were nowhere as accommodating, while more
conventional approaches had barred entry to geographers long ago.
Although it was rarely made explicit, a structuralist epistemology of one
sort or another (Harvey, for example, followed Piaget rather than
Althusser in *Social Justice and the City*) was infused almost subliminally
through the early development of Marxist geography.[7]

---

7. I am not claiming that early Marxist geography was exclusively structuralist or that
there was no awareness of the interpretive and political problems associated with a rigid
application of Althusserian doctrine. Unfortunately, when the new 'humanistic' geographers looked at Marxist geography, they often saw only its structural Marxism, blissfully
berating its allegedly unconscious extermination of the knowledgeable human agent without seeing that many Marxist geographers had already recognized the same problems. See,
for example, Duncan and Ley, 1982.

Two scales of analysis and theorization dominated the initial conjunction of Marxist political economy and critical human geography: the specifically urban, and the expansively international. In each, the fundamental approaches were similar. Urban geography and the geography of international development were examined as structured outcomes of the countervailing strategies of capitalist accumulation and class struggle, the generative and conflictful social processes shaping the production of space at every geographical scale.

Marxist geographers contributed significantly and centrally to the formation of an explicitly urban political economy built largely on 1) David Harvey's increasingly formalized Marxification of the 'urban process under capitalism'; 2) Manuel Castells's monumental adaptation of Althusser, Lefebvre, Alain Touraine (the French sociologist and theorist of 'social movements'), and the new school of French Marxist urban sociology; and 3) the constructive reactions to Harvey and Castells arising amongst radical political economists, geographers, and urban sociologists in Britain and North America.[8] The spatiality of the urban, the interaction between social processes and spatial forms, the possibility of a formative urban socio-spatial dialectic, were key issues of debate from the start and continue to be important in contemporary Marxist urban studies.

In an almost entirely separate sphere at first, the analysis of more global patterns of geographically uneven development, especially focused on Third World underdevelopment and dependency, produced another new and increasingly spatialized political economy of the international division of labour and the capitalist 'world system' of cores and peripheries. The most influential exponents of this neo-Marxist international political economy (for example, André Gunder Frank, Immanuel Wallerstein, Samir Amin, Arghiri Emmanuel, and the innovative group of Latin-American 'structuralists', as they were called) included very few Marxist geographers. Although the critical debates revolved around the inherently spatial structure of the international division of labour and what was clearly a worldwide process of geographically uneven development 'under capitalism' (to match Harvey's

---

8.  *Antipode* was the major journal of radical geographical research relating to the new urban political economy, and for a period in the 1970s had the highest circulation figures of any of the new radical journals in the social sciences. Also important as a vehicle for Marxist urban studies – with a marked sociological slant – was the *International Journal of Urban and Regional Research*, established in 1977 and based in Great Britain. After being centred for many years within the Graduate School of Geography at Clark University in Worcester, Massachusetts (the first issue appeared in 1969), *Antipode* is now published in Oxford by Basil Blackwell. For a convenient sampling of 'The Best of Antipode 1969–85', see the special double issue (Vol. 17, Numbers 2 and 3, 1985), the last published in the United States under the long and capable editorship of Richard Peet. See also Peet, 1977.

description of the urban process), the need to rethink the fundamental spatiality of capitalist development on a global scale was barely recognized. And when it was raised, the notion had little effect. The fundamental and immediate spatiality of urbanism was difficult to ignore (if not always easy to accept and understand), but the idea that the capitalist world economy was also presuppositionally spatial, the product of a similar spatialization process at another, less immediate scale, was much more elusive during the 1960s and 1970s.[9]

The new political economies of urbanization and international development attracted many adherents from geography and from the related fields of urban and regional planning.[10] But it was not long before the volatile mixture of perspectives that shaped these new political economies generated serious epistemological problems, especially with regard to the theorization of space and spatiality. Marxist geography teetered uncomfortably between the extremes of a pragmatic and anti-speculative historicism (which intrinsically rejected explicitly 'geographical' explanations of history and what many saw as an unacceptable emphasis on consumption and exchange relations versus relations of production); and a neo-Marxist structuralism (which seemed all too easily to breed determinisms, annihilate the politically conscious subject, and summarily to expel the theoretical primacy of historical explanation).

Adding Marxism to Modern Geography seemed a straightforward and worthy project. One could even set aside the larger epistemological conflicts as outside the project's domain. But a few geographers began to look in the other direction, apparently seeking to 'spatialize' historical

---

9. My own personal engagement with Marxist geography grew out of an attempt to respond to the radical critiques of my early work on the 'geography of modernization' in Africa (Soja, 1968, 1979; Soja and Tobin, 1974; Soja and Weaver, 1976). After exploring the sources of this critique, I began to think that I could do an even better job of radical reconstruction of the new 'development geography', as it came to be called. Chapter 4 reviews some of these first attempts at reconstruction.

10. It has become increasingly difficult to separate critical planning theory and debates on the nature of urban and regional planning practice from the unfolding encounter between Modern Geography and Western Marxism and the debates on the theorization of space. Indeed, the anglophonic tradition of planning education throughout the twentieth century has been one of the most important places for the preservation of practical geographical analysis, critical spatial theory, and the geographical imagination. Without its own distinctive disciplinary niche in the modern academic division of labour, urban and regional planning flexibly combined disciplinary traditions and emphases, providing, when not constrained by its own internal orthodoxies, an attractive environment for those seeking new, multi-disciplinary combinations and perspectives. Since the 1960s, following the expansion of the 'new social and political practice' that Lefebvre described as *spatial planning*, increasing numbers of Marxist geographers and spatial theorists, have become directly involved in urban and regional planning, exploring the state-managed production of space from inside the educational belly of the beast.

materialism and, albeit tentatively, to insert critical human geography into the interpretive core of the Western Marxist tradition. This was another matter altogether. Amidst great confusion, a second phase in the development of Marxist geography had begun – and it is still sorting itself out.

## The Provocative Inversion: Adding a Geography to Western Marxism

By the end of the 1970s, a fractious debate had arisen within Marxist geography over the difference that space makes to the materialist interpretation of history, the critique of capitalist development, and the politics of socialist reconstruction. Arguments spun back and forth between those who sought a more flexible and dialectical relation between space and society (Soja and Hadjimichalis, 1979; Soja, 1980; Peet, 1981); and those who saw in this effort a theoretical 'degeneracy', a dithering 'radical eclecticism', a politically dangerous and divisive spatial 'separatism' or 'fetishism' irreconcilable with class analysis and historical materialism itself (J. Anderson, 1980; Eliot Hurst, 1980; Smith, 1979, 1980, 1981). To some relatively sympathetic observers, Marxist geography appeared to be destroying itself from within, leading one to argue 'why geography cannot be Marxist' (Eyles, 1981) and another to lament what he saw as an irrational abandonment of spatial explanation in radical geographical analysis (Gregory, 1981).

A growing movement developed in Marxist geography and in urban and regional studies which appeared to be concluding that space and spatiality could fit into Marxism only as a reflective expression, a product of the more fundamental social relations of production and the aspatial (but nonetheless historical) 'laws of motion' of capital (Walker, 1978; Massey, 1978; Markusen, 1978). The geography of capitalism was a worthy subject, geographically uneven development was an interesting outcome of the history of capitalism, but neither was a *necessary* generative element in Marxist theory or a *presupposition* for historical materialist analysis. Marxism did not need a spatial 'fix', for it was not sufficiently broken.

This rekindling of Marxist orthodoxy was reinforced by a broader based 'critique of the critique' that was spreading through Western Marxism at about the same time, denouncing the theoretical inadequacies, overinterpretations, and depoliticizing abstractions of Althusserian structuralism and its neo-Marxist and 'Third Worldist' adherents. E.P. Thompson provided a symbolic banner for this attack in his long essay, *The Poverty of Theory* (1978), a passionately anti-structuralist reassertion of the primacy of history and British Marxist historicism, the diachronic against the synchronic, anglophonic concreteness against francophonic abstraction. Brenner (1977), an American Marxist historian, presented a parallel, if less passionate, critique of the neo-Marxist development theory of Wallerstein, Frank, and others. Here the revisionist error was more indelibly bourgeois than Stalinist (as Thompson labelled Althusser): a reversion to the value-less verities of Adam Smith. Marxist geographers were still invisible to these Western Marxist keepers of the faith, but the repercussions of their critiques were nevertheless deeply felt in both urban and international political economy, the former through a growing aversion to Castellsian urban structuralism and the latter through a formidable resistance to assigning a more central, structuring role to geographically uneven development in the history of capitalism.

Despite the power of this reassertive historicist orthodoxy, the project of spatializing Marxist theory remained alive. Its major means of survival was an increasingly well-articulated theoretical argument that the organization of space was not only a social product but simultaneously rebounded back to shape social relations. Hoping to ward off the rising orthodoxy from within (in part by appealing respectfully to Lefebvre), I tried to capture this two-way flow through the assertion of a 'socio-spatial dialectic' and the need for a radical 'spatial praxis' in what I initially called a new 'topian' Marxism (see Chapters 3 and 4). Taking another, perhaps broader view, Derek Gregory in *Ideology, Science and Human Geography* (1978), probably the decade's most insightful and comprehensive reinterpretation of geographical explanation, put the argument this way:

> The analysis of spatial structure is not derivative and secondary to the analysis of social structure, as the structuralist problematic would suggest: rather, each requires the other. Spatial structure is not, therefore, merely the arena within which class conflicts express themselves (Scott 1976, 104) but also the domain within which – and, in part, through which – class relations are constituted, and its concepts must have a place in the construction of the concepts of determinate social formations ... spatial structures cannot be *theorized* without social structures, *and vice versa*, and ... social structures cannot be *practised* without spatial structures, *and vice versa*. (Gregory, 1978, 120–1)

Gregory argued for a more committed and emancipatory form of geographical explanation that would draw upon both structural and 'reflexive' (phenomenological, hermeneutic) epistemologies to give human geography a 'place' among the critical social sciences.

The challenge was clear. There was a complex and problem-filled interaction between the production of human geographies and the constitution of social relations and practices which needed to be recognized and opened up to theoretical and political interpretation. This could not be done by continuing to see human geography only as a reflection of social processes. The created spatiality of social life had to be seen as simultaneously contingent and conditioning, as both an outcome and a medium for the making of history – in other words, as part of a historical and geographical materialism rather than just a historical materialism applied to geographical questions.

Even David Harvey, in the middle of a decade of dedicated Marxification, occasionally reversed himself to begin spatializing Marxism in his early formulations of capital's search for a 'spatial fix' (which he perceptively traced back to the older spatializations of Hegel and Von Thunen); and, especially, his descriptions of the tense interplay between preservation and destruction in the urban built environment (or what he would later call the 'restless formation and reformation of geographical landscapes'). Harvey remained ambivalent, taking a few hesitant and imaginative leaps toward the socio-spatial dialectic (Harvey, 1977, 1978); but always seeming to return to the formalism of rigorous Marxification, even as the limits to this formalizing became more vividly apparent. Some of Harvey's best students (Richard Walker and Neil Smith, for example) formed a harder line than did Harvey himself, leading the charge against the 'provocative inversion' and its proposed inter-contingency of space and class, spatiality and society.

The resistance, however, was much more widespread within Modern Geography and Western Marxism. The former habitually avoided making the spatial contingency of society and behaviour too explicit, except via the 'neutral' physical force of the friction of distance, Modern Geography's depoliticizing shroud. Instead, this possibility was preserved almost as an in-house secret to be shared by adherents but not to be publicly revealed. Spatial contingency smacked too sharply of the errant environmentalism of an embarrassing past and clashed with the ascribed position of geography within the Modern academic division of labour, innocently descriptive of areal differentiation but sworn never again to claim any geographical determination of the social, at least not within earshot of the assiduously social sciences.

Most Marxists, especially at a time of rising orthodoxy, could perceive in the asserted spatial contingency of class only another attempt to impose an 'external' constraint upon the freedoms of class consciousness and social will in the making of history. In addition, the theoretical arguments behind this spatial contingency seemed to spin off too haphazardly from the eclectic structuralism and overassertiveness of neo-Marxist political economies of urbanization and international development. Equally important, the provocative inversion had not yet been backed by a more formal theoretical and epistemological elaboration or by a body of cogent empirical analysis that made the necessary political connections. For the mainstream of Marxist geography, therefore, the new dialectic was tempting but had not yet demonstrated sufficiently its use value.

It is no wonder then that well into the 1980s Marxist geography seemed to be dancing a tedious gavotte around the materialist interpretation of spatiality, moving forward only to glide back again. Between Harvey's *Social Justice and the City* (1973) and *The Limits to Capital* (1982), there were very few original and synthesizing books produced by individual Marxist geographers and almost nothing at all written in English that systematically and explicitly advanced the call for a historico-geographical materialism. There were several important editorial collections which presented a radical spatial perspective (Peet, 1977; Dear and Scott, 1981; Carney, Hudson, and Lewis, 1980), but, in a sign of what was to come, some of the most influential critical human geographies written during this period were written by geographers outside (but looking into) the Marxist mainstream (Olsson, 1980; Scott, 1980; Brookfield, 1975; and, above all, Gregory, 1978).

With a few exceptions, Marxist geography seemed to be spinning and turning into the 1980s to avoid what would eventually become insistently obvious: that Marxism itself had to be critically restructured to incorporate a salient and central spatial dimension. The inherited orthodoxies of historical materialism left almost as little room for space as the rigid cocoons of bourgeois social science. Making geography Marxist was thus not enough. Another and much more disruptive round of critical thinking was necessary to spatialize Marxism, to recombine the making of history with the making of geography. The provocative inversion had to be turned into a third critique, a critique of the critique of the critique that initiated the development of Marxist geography.

For the most part, the impetus for this deconstructive and reconstitutive third critique originated outside Marxist geography and was carried forward by critical scholars who were often oblivious to the existence and accomplishments of Marxist geographers. The wider debate responded primarily to the perceived peculiarities of 'late capitalism', in particular the perplexing societal restructuring that seemed to be shattering long-established political, economic, cultural, ideological, and intellectual patterns. New combinations appeared to be emerging which defied conventional interpretation and political practice left, right, and centre. The search for an appropriate theoretical and political response to the restructuring of capitalism brought a small but highly diverse group of leftist critical scholars to a remarkably similar conclusion: that the formative spatiality of social life had become a particularly crucial and revealing interpretive window onto the contemporary scene but that the spatial viewpoint has been obscured by a long heritage of neglect and mystification. With this realization, repeated by scholar after scholar from many different specialized fields, the debate on the political theorization of space became more generalized than it had ever been before, ushering in another, less parochial, phase in the encounter between Modern Geography and Western Marxism.

## Passages to Postmodernity: the Reconstitution of Critical Human Geography

Bringing up to date the development of Marxist geography and its attendant re-theorization of the historical geography of capitalism is a necessarily eclectic exercise. It can now be said that geographical modes of analysis are more centrally attached to contemporary political and theoretical debates than at any other time in this century. But the attachment derives from many different sources, takes a variety of forms, and resists easy synthesis. It is also still very tentative and limited in its impact, for the spatialization of critical theory and the construction of a new historico-geographical materialism have only just begun and their initial impact has been highly disruptive, especially for the two modernist traditions that have shaped the development of Marxist geography over the past twenty years.

Just as contemporary Western Marxism seems to have exploded into a heterogeneous constellation of often cross-purposeful perspectives, Modern Geography has also started to come apart at its seams, unravelling internally and in its old school ties with the other nineteenth-century disciplines that defined the modern academic division of labour. The grip of older categories, boundaries, and separations is weakening. What was central is now being pushed to the margins, while the once tactful fringes boldly assert a new-found centrality. The shifting, almost kaleidoscopic, intellectual terrain has become extremely difficult to map for it no longer appears with its familiar, time-worn contours.

This unsettled and unsettling geography is, I suggest, part of the postmodern condition, a contemporary crisis filled, like the Chinese pictograph for crisis and Berman's vaporous description of modernity in transition, with perils and new possibilities; filled with the simultaneous shock of the old and the new. To bring back an earlier argument, another culture of time and space seems to be taking shape in this contemporary context and it is redefining the nature and experience of everyday life in the modern world – and along with it the whole fabric of social theory. I would locate the onset of this passage to postmodernity in the late 1960s and the series of explosive events which together marked the end of the long post-war boom in the capitalist world economy. And to identify the most insightful early cartographers of this portentous transition, I would turn again to the prefigurative writings of Lefebvre and Foucault, Berger and Mandel, for it is there that the geography of postmodernization was most acutely perceived. Although the links between them are not always direct and intentional, the intellectual trajectories of these four foundational postmodern geographers intersect in the contemporary deconstruction and reconstitution of modernity. I will use them once more to help explore the variegated postmodern landscapes of critical human geography.

## The convergence of three spatializations

Lefebvre, Foucault, Berger and Mandel all crystallized their assertions of the significance of spatiality at a crucial historical moment, when the most severe global economic crisis since the Great Depression had signalled to the world the end of the post-war boom and the onset of a profound restructuring that would reach into every sphere of social life and shatter the conventional wisdoms built upon simplistic projections from the immediate past. Although Mandel is not quite so explicit, the other three clearly rotate their arguments around the realization that it is now space more than time that hides things from us, that the demystification of spatiality and its veiled instrumentality of power is the key to making practical, political, and theoretical sense of the contemporary era.

Juxtaposing these arguments sets in motion a creative convergence between three different paths of spatialization that I shall call 'post-historicism', 'postfordism', and 'postmodernism'. The first of these spatializations is rooted in a fundamental reformulation of the nature and conceptualization of social being, an essentially ontological struggle to rebalance the interpretable interplay between history, geography, and society. Here the reassertion of space arises against the grain of an onto-logical historicism that has privileged the separate constitution of being in time for at least the past century.

The second spatialization is directly attached to the political economy of the material world and, more specifically, to the 'fourth moderni-zation' of capitalism, the most recent phase of far-reaching socio-spatial restructuring that has followed the end of the long post-war economic boom. The term 'postfordism' is tentatively chosen to characterize the transition from the regime of accumulation and mode of regulation that consolidated after the Great Depression around large-scale, vertically-integrated, industrial production systems; mass consumerism and sprawling suburbanization; the centralized Keynesian planning of the welfare state; and increasing corporate oligopoly.[11] Here too it can be argued that space makes a critical difference, that revealing how spatial restructuring hides consequences from us is the key to making political and theoretical sense of the changing political economy of the contemporary world.

The third spatialization is couched in a cultural and ideological reconfiguration, a changing definition of the experiential meaning of modernity, the emergence of a new, postmodern culture of space and time. It is attuned to changes in the way we think about and respond to the particularities – the perils and possibilities – of the contemporary moment via science, art, philosophy, and programmes for political action. Postmodernism overlaps with posthistoricism and postfordism as a theoretical discourse and a periodizing concept in which geography increasingly matters as a vantage point of critical insight.

The confluence of these three spatializations is effectively exemplified in the recent work of Fredric Jameson, perhaps the pre-eminent American Marxist literary critic. In an essay entitled 'Postmodernism, or the Cultural Logic of Late Capitalism', Jameson captures the spatial specificity of the contemporary *Zeitgeist.*

> Postmodern (or multinational) space is not merely a cultural ideology or fantasy, but has genuine historical (and socio-economic) reality as a third great original expansion of capitalism around the globe (after the earlier expansions of the national market and the older imperialist system, which each had their own cultural specificity and generated new types of space appropriate to their dynamics).... We cannot [therefore] return to aesthetic practices elaborated on the basis of historical situations and dilemmas which are no longer ours ... the conception of space that has been developed here suggests that a model of political culture appropriate to our own situation will necessarily have to raise spatial issues as its fundamental organizing concern. (Jameson, 1984, 88–9)

Jameson derives much of his conceptualization of space from the *Raumgeist* of Lefebvre, whom he helped to reach a larger American audience in the early 1980s. But there are other echoes to be heard. Jameson provisionally defines the spatialized model of radical political culture appropriate to the contemporary (postmodern) situation as an 'aesthetic of cognitive mapping', an ability to see in the cultural logic and

---

11. The most widely used term to describe this recent reconfiguration of capitalism has been 'post-industrial'. The term has its appeal, but misdirects our attention away from the continuing centrality of industrial production and the labour process in the contemporary restructuring of capitalist societies. It is as absurd in its way as describing what has been happening as 'post-capitalism' or 'the end of ideology'.

forms of postmodernism an instrumental cartography of power and
social control; in other words, a more acute way of seeing how space
hides consequences from us. He refers specifically to the work of Kevin
Lynch (1960) on 'images of the city', but the insinuating connections
link back not only to Lefebvre and Berger but also to Foucault.
Foucault's 'carceral city' of cells, ranks, and enclosures, for example, is
transposed by Jameson on to the landscape of perhaps the quintessential
postmodern place, Los Angeles, the production site for some of the
most pervasive and persuasive cognitive imagery in the world today.[12]
Jameson maps out from Los Angeles and other postmodern landscapes,
material and literary, a hidden and insidious human geography that must
become the target for a radical and postmodern politics of resistance, a
means of tearing off the gratuitous veils that have been drawn over the
instrumentality of contemporary restructuring processes.

Foucault's own explicit but more deflected emphasis on spatiality has
already been discussed. His archeology and genealogy of knowledge
provided an important passageway to the postmodern cultural critique
of spatiality and the cartography of power. Instead of abandoning
radical politics, as Anderson and others have claimed he did, Foucault
added his voice to Lefebvre's in spatializing the political project of the
Left. 'The real political task in a society such as ours', Foucault writes,
'is to criticize the working of institutions which appear to be both neutral
and independent; to criticize them in such a manner that the political
violence which has always exercised itself obscurely through them will be
unmasked, so that one can fight them' (see Rabinow, 1984, 6).[13] The
'fight' extends beyond the exploitative institutions of capitalism alone to
all 'disciplinary technologies' wherever they are found, even in the
realms of existing socialisms. This delineation of spatial struggle is but
one step away from Jameson's postmodernism of resistance. It is an
expressly geopolitical strategy in which spatial issues are the fundamental organizing concern, for disciplinary power proceeds primarily
through the organization, enclosure, and control of individuals in
space.[14]

Jameson's use of postmodernism as a periodizing concept brings us to
the Mandelian connection in the contemporary reassertion of space in

---

12. In 1984, Jameson, Lefebvre, and I took a spiralling tour around the centre of Los
Angeles, starting at the Bonaventure Hotel. I try to recapture our travels in chapter 9.

13. Despite a very sympathetic treatment of Foucault's spatializations, Rabinow
continues to describe Foucault's work, misleadingly I would argue, as 'a form of critical
historicism'.

14. The spatial analytics of Foucault are being recaptured with particular *élan* by
Derek Gregory and his students at the University of Cambridge. See Gregory, forthcoming.

critical social theory. Jameson draws directly from Mandel's interpretation of *Late Capitalism* (Mandel, 1975) and the changing role of geographically uneven development, in which Mandel makes the key conjunction between periodization and spatialization in the macro-political economy of capitalism. The 'long wave' periodicity of crisis and restructuring defines not only a series of historical eras but simultaneously a sequence of spatialities, a changing regional configuration of 'uneven and combined' capitalist development that can be mapped on to the sequence of modernities discussed in the preceding chapter. This spatio-temporal patterning plays a significant role in Jameson's critique of reactionary postmodernism and in a wide range of contemporary studies of postfordist economic restructuring. It also provides a useful framework through which to examine the changing urban, regional, and international geography of capitalism – what contemporary analysts have begun to describe as capitalism's multi-layered spatial divisions of labour (Massey, 1984).

There is still one more important passage to postmodernity to be explored, one more arena in which the three spatializations have begun to converge. This is the postmodern deconstruction and reconstitution of Marxist geography itself, a story which will be told again, in many different ways, in every subsequent chapter. Here, I intend only to introduce and outline a few of the most important contemporary debates and developments.

## The postmodernization of Marxist geography

Marxist geography has changed significantly in the 1980s. Its boundaries have broadened and become more flexibly defined so that its influence extends more deeply into the realms of critical social theory than ever before. At the same time, its centre of gravity has become more difficult to locate except as part of a larger project of insistent spatialization, wherein traditional Marxist categories and definitions of geography are becoming significantly disassembled and rearranged. Adding to the confusion, a new zest for the empirical has turned many Marxist geographers away from open theoretical debate just when the debate has reached a much larger – and more critical – audience. The passage to postmodernity thus seems to be having a distinctively disintegrating effect.

This disintegration should come as no surprise, especially to those attuned to the influential writings of David Harvey, who, more than any other geographer, has shaped and continues to shape the course of Marxist geographical inquiry. As Harvey notes over and over again:

The insertion of concepts of space and space relations, of place, locale, and milieu, into any of the various supposedly powerful but spaceless social theoretical formulations has the awkward habit of paralyzing that theory's central propositions ... Whenever social theorists actively interrogate the meaning of geographical and spatial categories, either they are forced to so many ad hoc adjustments that their theory splinters into incoherence or they are forced to rework very basic propositions. (1985b, xiii)

Harvey also links, in his more recent work at least, the disruptive effects of spatialization with the rigidities of historicism in Western Marxism as well as in the liberal social sciences.

Marx, Marshall, Weber and Durkheim all have this in common: they prioritize time and history over space and geography and, where they treat of the latter at all, tend to view them unproblematically as the stable context or site for historical action.... The way in which space relations and the geographical configurations are produced in the first place passes, for the most part, unremarked, ignored.... Marx frequently admits of the significance of space and place in his writings ... [but] geographical variation is excluded as an 'unnecessary complication'. His political vision and his theory are, I conclude, undermined by his failure to build a systematic and distinctively geographical and spatial dimension into his thought. (1985c, 141–3)

These passages mark a significant shift of emphasis in Harvey's work and highlight many of the dilemmas currently facing Marxist geography and geographers.

In *Limits to Capital* (1982), Harvey reached out from the heart of Marxist geography to the wider realm of Western Marxism and modern critical social theory, presenting a demonstrative argument for a spatialized Marxism and a spatialized critique of capitalist development. Giving a concrete geography to *Capital* and to capitalism, however, was at once a major *tour de force* and an invitation to theoretical paralysis. *Limits* thus combined, incongruously, both the crowning achievement of formalistic Marxist geography and the opening salvoes for the necessary deconstruction and reconstitution of this very achievement. Recognizing this stressful ambivalence and the possibility that he might have to pull the tightly woven rug from under himself, Harvey almost abandoned the project. Encouraged by ex-students and other supporters, however, Harvey completed his masterful synthesis hoping, it would seem, that the disruptive impact of 'inserting' space would not diminish the force of his avowedly historical and geographical materialism.

Looking back at the critical response to *Limits*, Harvey expressed his concern that most readers seemed to be somewhat confused over the message it was presenting. In the preface to *Consciousness and the Urban Experience*, he notes:

> Curiously, most reviewers passed by (mainly, I suspect, out of pure disciplinary prejudice) what I thought to be the most singular contribution of that work – the integration of the production of space and spatial configurations as an active element within the core of Marxist theorizing. That was the key theoretical innovation that allowed me to shift from thinking about history to historical geography and so to open the way to theorizing about the urban process as an active moment in the historical geography of class struggle and capital accumulation. (1985b, xii)

Most Marxist geographers got the message, but they were already convinced and many had begun their own turnarounds from earlier positions antagonistic to the provocative spatialization of Western Marxism. In her preface to *Spatial Divisions of Labour: Social Structures and the Geography of Production* (1984, x), Doreen Massey writes:

> My basic aim had been to link the geography of industry and employment to the wider, and underlying structures of society ... The initial intention, in other words, was to start from the characteristics of economy and society, and proceed to explain their geography. But the more I got involved in the subject, the more it seemed that the process was not just one way. It is also the case – I would argue – that understanding geographical organisation is fundamental to understanding economy and society. The geography of society makes a difference in the way it works.
>
> If this is true analytically, it is also true politically. For there to be any hope of altering the fundamentally unequal geography of British economy and society (and that of other capitalist countries, too), a politics is necessary which links questions of geographical distribution to those of social and economic organisation.

Massey takes her new path into an analysis of the locational particularities of 'spatial structures of production', seeking to open a middle ground of regional political economy freed from both the iron determination of capitalism's laws of motion and the inane indeterminacy of geographical empiricism.[15]

Neil Smith also strays cautiously off the path of Marxist orthodoxy to

---

15. Massey's work has inspired a new obsession with 'localities' in British geographical research, a privileging of the particularities of place that has yet to prove particularly fruitful. The emphasis on localities has in turn engendered a growing debate on the theoretical implications of what is described as radical geography's 'empirical turn'. See the many recent articles on this debate in *Antipode* (Cooke, 1987; Smith, 1987; Cochrane, 1987; Gregson, 1987).

make a similar argument in the preface to his *Uneven Development: Nature, Capital and the Production of Space* (1984, xi):

> Occupying the common ground between the geographical and political traditions, a theory of uneven development provides the major key to determining what characterizes the specific geography of capitalism.... But one cannot probe too far into the logic of uneven development without realizing that something more profound is at stake. It is not just a question of what capitalism does to geography but rather of what geography can do for capitalism.... From the Marxist point of view, therefore, it is not just a question of extending the depth and jurisdiction of Marxist theory, but of pioneering a whole new facet of explanation concerning the survival of capitalism in the twentieth century.... Geographical space is on the economic and political agenda as never before. The idea of the 'geographical pivot of history' takes on a more modern and more profound meaning than Mackinder could have imagined.

The prefatory admissions of Harvey, Massey, and Smith are not always followed to their appropriate denouement in the subsequent texts, for each is hesitant to engage too deeply in the necessarily transformative deconstruction of historical materialism and its despatializing master narratives. But while historicism is shielded from a rigorous and systematic critique, there is a new confidence regarding the theoretical and political significance of space. The need to justify and defend the theoretical assertion of a historical and geographical materialism is much less pressing than it once was. The time has come instead to demonstrate its political and empirical power through the analysis of the 'specific geography of capitalism':

> Geographical space is always the realm of the 'concrete and the particular'. Is it possible to construct a theory of the concrete and the particular in the context of the universal and abstract determinations of Marx's theory of capitalist accumulation? This is the fundamental question to be resolved. (Harvey, 1985c, 144)

But as Mills stated, every cobbler thinks leather is the only thing. The outreach of this more theoretically confident and assertive Marxist geography is still both insufficiently comprehensible and uncomfortably threatening to the modern academic division of labour, with its reified disciplinary compartments and intellectual territoriality. Moreover, the attack is being launched from what many still perceive as a minor disciplinary backwater into the still sanctified domains of the historical imagination. Retaining the *imprimatur* of Marxism is enough to frighten the contemporary FRUMPs (formerly radical upwardly mobile professionals, a term Harvey uses in a recent polemical defence of his position – see Harvey et al., 1987), but to have the theoretical and empirical trails blazed by 'uppity' geographers has been more than the established market could bear. In the 1980s, even some of the most sympathetic and open-minded social analysts have begun to lash back at the determined space invaders, covering their disciplinary flanks against the disruptive effects of the 'postmodern' spatialization of critical social theory and analysis.

The most immediate and direct response to the intrusions of Marxist geographers has come from sociologists who continue to assume that they are in control of the spatialization of social theory, as indeed they largely have been, by default, since the late nineteenth century. Radical political economists have listened attentively to Harvey and the new Marxist geography but tend to keep their calculated distance, incorporating only the bare essentials of a spatialized Marxism. Marxist historians, when they recognize what is happening at all, typically respond by extending their best wishes (while tacitly assuming that radical historiography has already done what was necessary concerning geography years ago). But Marxist and radical Weberian sociologists have been more deeply involved in the spatializing project from the beginning and could not simply set it aside once it started to have a paralysing effect on their most cherished sociological propositions and principles. The spatialization of social theory, many felt, was going too far and needed to be appropriately disciplined.

The most prominent figure leading this sociological backlash has been Peter Saunders, whose work *Social Theory and the Urban Question* (1981, second edn 1986) provides an excellent and comprehensive overview of the historical development of urban social and spatial theory. David Harvey, immediately after observing the disintegrative impact of spatialization on established theoretical propositions in the social sciences (see above), turns to Saunders's first edition of *Social Theory and the Urban Question* to exemplify his argument.

> Small wonder, then, that Saunders (1981, 278), in a recent attempt to save the supposed subdiscipline of urban sociology from such an ugly fate, offers the extraordinary proposition, for which no justification is or ever could be found, that 'the problems of space ... must be severed from concern with specific social processes'.

Saunders is even more emphatic in the second edition, after having carefully surveyed the recent work of Marxist geographers. Summing up a chapter aimed explicitly at defining 'A non-spatial urban sociology', Saunders writes:

Ever since the work of Robert Park early in this century, urban sociologists have been developing theoretical insights which have been undermined by the insistent attempt to mould them to a concern with space. It is time to rid ourselves of this theoretical straitjacket. It is time to put space in its place as a contingent factor to be addressed in empirical investigations rather than as an essential factor to be theorized in terms of its generalities. It is time for urban social theory to develop a distinctive focus on some aspect of social organization in space rather than attempting to sustain a futile emphasis on spatial organization in society. It is time, in short, to develop a non-spatial urban sociology which, while recognizing the empirical significance of spatial arrangements, does not seek to elevate these arrangements to the status of a distinct theoretical object. (1986, 287–88)

Saunders rides fitfully to the rescue, in the nick of time, to put space in its same old place and to reify once more the traditional domain of Modern Sociology. In stripping urban sociology of a theoretical object that has anything to do with space or, for that matter, with the city, Saunders comes perilously close to vaporizing urban social theory entirely. In the end, he slips backward into a 'sociology of consumption' as the definitive focus of theoretical and substantive concern, retaining the adjective 'urban' only as 'a matter of convention', a useful way to 'maintain the intellectual continuity of the field' (Ibid., 289).

The debate on the 'specificity of the urban', that is to say, whether specifically urban social and spatial forms can provide an appropriate object for theorization, has always been a source of confusion and disagreement within the coalition of geographers, sociologists, and political economists that had formed in the 1970s to develop a new critical interpretation of capitalist urbanization. Much of the confusion arose from the equivocal conceptualization of the 'urban question' presented by Manuel Castells, the coalition's most influential Marxist sociologist. On the one hand, Castells attacked the overspecifications of the urban from the Chicago School to its alleged 'left-wing' extension in the works of Henri Lefebvre, arguing that there was no specifically urban problematic. Seeing urbanism as a distinctive 'way of life' was an ideological smokescreen obscuring larger societal problems that are expressed in cities but are not confined, epistemologically and politically, to the urban context. On the other hand, Castells conveniently respecified the urban as a theoretical object by focusing on the urban politics of collective consumption and the mobilization of distinctively urban social movements. While Castells would eventually extract himself from this epistemological trap, Saunders falls back into it in a desperate effort to maintain the 'intellectual continuity' and nominal disciplinary integrity of urban sociology.

The debate on the specificity of the urban was more than an exercise in epistemological gymnastics. From the beginning, it was a disciplinary conflict between radical sociology and Marxist geography over the spatialization of social theory, over just how far the reassertion of space would be allowed to go. Castells's Althusserian concoction of the urban question deflected the bolder assertions of Lefebvre, who, far from fetishizing the urban, was developing a more general argument that social struggle in the contemporary world, be it urban or otherwise, was inherently a struggle over the social production of space, a potentially revolutionary response to the instrumentality and uneven development of the specific geography of capitalism. In other words, the urban social movements and struggles over collective consumption that had become so central to what I am tempted to call 'Late Modern' radical sociology were being seen as part of a larger spatial problematic in capitalist development. (I will retrace these earlier arguments again in the next two chapters.)

In the 1980s, however, there were new twists being given to these older debates. Marxist geographers such as Harvey, Smith, and others cut through their former ambivalence to join together in developing a transformative historical-geographical materialism, a much more radical project than the earlier call for a spatialized urban political economy. The project has been supported and sustained, as I have previously described, by a host of 'outsiders' as the debate on the significance of spatiality in social theory and social practice became more widespread than ever before. More significantly for the current discussion, new voices were heard from within Modern Sociology calling loudly for the insertion of space at the very heart of social theory – for an even greater spatialization than had been achieved through the 1970s. The pesky geographers could still be brushed off as obsessive fetishizers, infatuated with their own 'leather'. Sociologists such as Anthony Giddens, John Urry, and a spatially reawakened Manuel Castells were another matter.

Castells's more recent work is marked by two apparent reversals if seen against the conventional portrayal of his earlier contributions to anglophonic urban sociology. The first comes from a softening of his stance against Henri Lefebvre and a greater willingness to accept the importance of an assertively spatial problematic in the interpretation of urban politics and sociology. The ever-slippery Castells never quite completes this reversal, but the following passage from *The City and the Grass Roots* (1983, 4) suggests a greater willingness to accommodate the Lefebvrean project:

Space is not a 'reflection of society', it *is* society.... Therefore, spatial forms, at least on our planet, will be produced, as all other objects are, by human action. They will express and perform the interests of the dominant class according to a given mode of production and to a specific mode of development. They will express and implement the power relationships of the state in an historically defined society. They will be realised and shaped by the process of gender domination and by state-enforced family life. At the same time, spatial forms will be earmarked by the resistance from exploited classes, from oppressed subjects, and from dominated women. And the work of such a contradictory historical process on the space will be accomplished on an already inherited spatial form, the product of former history and the support of new interests, projects, protests, and dreams. Finally, from time to time, social movements will arise to challenge the meaning of spatial structure and therefore attempt new functions and new forms.

Castells falls short of the postmodern proclamation that it is now space more than time, geography more than history, that hides consequences from us. But he seems at least more open to the possibility than he once had been.

Castells's second reversal is also not a complete about-face, but runs equally against the grain of conventional portrayals of his contributions to urban studies. It arises from a reinvigorated interest in industrial production and technology and its effects on the urbanization process. Here Castells is part of a much larger shift of attention in contemporary urban and regional studies and in Marxist geography, that is not so much a denial of the importance of collective consumption issues as a recognition that the dynamics of industrial production and restructuring must be understood first, before we can make theoretical and practical sense of the politics and sociology of consumption. Castells might argue that this is what he has been saying all along, reminding those who criticized his supposed 'consumptionism' of his early works on urban industrialization (for example, Castells and Godard, 1974). But he is currently pursuing these interests with renewed enthusiasm and insight (see Castells, 1985), alongside a growing group of postfordist, if not also postmodern, industrial and production-oriented urban geographers (see, for example, Scott and Storper, 1986).

Saunders tries hard to deflect the provocative spatial turns of Giddens and Urry, the apparent reversals of Castells, and the emboldened attempts by Marxist geographers to create an historical and geographical materialism, by appealing to the 'theoretical realist' philosophy of social science that Urry helped to develop (see Keat and Urry, 1982), Giddens draws heavily upon, and the geographer, Andrew Sayer (1984), has so carefully codified. Saunders uses his newfound theoretical realism to 'put space in its place' as merely a contingent factor to be addressed in empirical investigations rather than an essential part of social theorization. In doing so, he attaches the sociological backlash to a larger philosophical and methodological debate that has confusingly shaped the postmodernization of Marxist geography.

The impact of theoretical realism (Bhaskar, 1975, 1979; Harre, 1970; Harre and Madden, 1975) on the reassertion of space in social theory has been many-sided and far-reaching. With its flexible synthesis of structuralism and hermeneutics, its insistence on situating social theory and social practice in the conjunctural effects of time and space, and its adaptation of a Marxian notion of praxis whilst simultaneously subjecting Marxism to a vigorous 'contemporary' critique, theoretical realism seemed to provide an almost ideal epistemological framework for postmodern critical human geography. If it did not exist, it would have had to be invented! But it has filtered into the spatial discourse of the 1980s with a disruptive ambivalence, both helping and hindering the development of critical spatial theory.

Realist philosophy over the past decade has inspired the most systematic, forceful, and influential assertions of the significance of space in the construction of social theory, primarily via the structuration theory of Anthony Giddens. It has also provoked, largely through the work of Sayer (1979, 1982, 1984, 1985), a constrictive countercurrent which argues that Marx, Weber, Durkheim, and others, may have been justified in paying so little attention to space in their abstract theoretical work because 'the difference that space makes' is important only at the level of the concrete and empirical. The proper path for (post-Marxist?) geography to take is thus primarily an empirical one, leaving behind the grander theoretical debates, whether set in the mould of space-time structuration or historical-geographical materialism. More than any other event, this realist counter-assertion – propelled back to the empirical drawing-board by the perplexities for the left arising from the regressive victories of Ronald Reagan and Margaret Thatcher – helped to split the new Marxist geography consensus that had been emerging in the 1980s.

One of the effects of this disintegrative implosion has been to encourage the reactionary sociological backlash against Marxist geography and to move many erstwhile Marxist geographers to join in with plaintive '*mea culpas*'. Another effect has been a strategic Marxist retrenchment, led by David Harvey, to keep alive the project of historical-geographical materialism against the rising anti-theoretical (and frequently anti-Marxist) onslaught. Confrontations between these two rigidifying positions, complicated by growing confusion over exactly who is on what side, continue to fill the pages of *Antipode* and *Society and Space* and absorb perhaps too much of the energies of those involved.[16]

---

16. See especially the molehill of embittered commentaries instigated by Saunders and Williams (1986), a crude radical-baiting attack on the alleged Marxist-realist orthodoxy that the authors claim has captured British urban and regional studies. David Harvey's

Fortunately, there has recently also begun to develop the glimmerings of a more reconstructive postmodern critical human geography arising from these still ongoing 'Late Modern'[17] confrontations. It continues to draw inspiration from the emancipatory rationality of Western Marxism but can no longer be confined within its contours, just as it cannot be constrained by the boundaries of Modern Geography. It can perhaps best be described as a flexible specialization, to adopt a term from current research on postfordist industrial organization and technology. Flexible specialization in the workplace of critical human geography means a resistance to paradigmatic closure and rigidly categorical thinking; the capacity to combine creatively what in the past was considered to be antithetical/uncombinable; the rejection of totalizing 'deep logics' that blinker our ways of seeing; the search for new ways to interpret the empirical world and tear away its layers of ideological mystification. It thus involves a temporary suspension of epistemological formalism to allow the new combinations of history and geography to take shape dialectically and pragmatically, unburdened by the biases of the past but guided nonetheless by the testing ground of praxis.

This emerging postmodern critical human geography must continue to be built upon a radical deconstruction, a deeper exploration of those critical silences in the texts, narratives, and intellectual landscapes of the past, an attempt to reinscribe and resituate the meaning and significance of space in history and in historical materialism. Spatial deconstruction aims to 'reverse the imposing tapestry' of the past, to use Terry Eagleton's words (Eagleton, 1986, 80), exposing the dishevelled tangle of threads that constitutes the intellectual history of critical social thought. This task has only just begun and is already meeting with fierce resistance, especially from those identified by Foucault as 'the pious descendants of time'. Spatial deconstruction must therefore also be sufficiently flexible to parry the reactionary thrusts of historicism and avoid the simplistic defence of anti-history or, even worse, a new and equally obfuscating spatialism. The objective is, after all, a politically charged historical geography, a spatio-temporal perspective on society and social life, not the resurrection of geographical determinism.

---

personal response (to what was indeed a personal attack), along with a series of largely self-serving reactions to the bubbling confusion of misrepresentations and shadow-boxing feints and left-crosses, has recently been published in *Society and Space* (Harvey et al., 1987) under the presumptuous title: 'Reconsidering social theory: a debate'.

17. I use the term 'Late Modern' to refer to a continued defence of the modern intellectual division of labour – including Western Marxism and Modern Geography – but with a few contemporary and adaptive twists, such as recognizing the impact of postfordist and postmodern restructuring processes. A brilliant example of this flexible halfway house of Late Modern Marxist geography is Harvey's recent paper, 'Flexible Accumulation Through Urbanization: Reflections on "Post-Modernism" in the American City' (1987).

Deconstruction alone is not enough, however, no matter how effect-ively the critical silences are exposed. It must be accompanied by an at least tentative reconstruction grounded in the political and theoretical demands of the contemporary world and able to encompass all the scales of modern power, from the grand strategies of global geopolitics to the 'little tactics of the habitat', to again borrow from Foucault. This reconstituted critical human geography must be attuned to the emanci-patory struggles of all those who are peripheralized and oppressed by the specific geography of capitalism (and existing socialism as well) – exploited workers, tyrannized peoples, dominated women. And it must be especially attuned to the particularities of contemporary restructuring processes and emerging regimes of 'flexible' accumulation and social regulation, not merely to display a newfound empirical prowess but to contribute to a radical postmodernism of resistance.

Flexible specialization is again a necessary accompaniment to this strategic reconstruction of critical human geography, whether it be focused on the interpretation of the new technology and restructured organizational forms of the postfordist political economy, the cultural logic of postmodernism in art and ideology, or the ontological struggles of a posthistoricist critical theory. These three pathways of spatialization and potentially radical spatial praxis must be combined as compatible, not competitive, fields and viewpoints. Similarly, the new zest for the empirical, even under the pretence of its political practicality, must not close off theoretical debate and discussion, for there is nothing so prac-tical as good spatial theory.

Flexible deconstruction and reconstitution will not be easy, for it must contend not only with a continuing 'Late Modern' resistance carrying with it the privileged baggage of the past, but must also deal with a rising neo-conservative postmodernism flexing its muscles in the present, monopolizing the debate on what now must be done to meet the challenges of a new modernity. Neo-conservative postmodernism is using deconstruction to draw even more obfuscating veils over the instrumentality of restructuring and spatialization, reducing both history and geography to meaningless whimsy and pastiche (or to mere 'factual-ity' again) in an effort to celebrate the postmodern as the best of all possible worlds. Opposition to restructuring is made to appear as extremism, the very hope of resistance becomes tinged with the absurd. Marxism is equated only with totalitarianism; radical feminism becomes the destruction of the family; the anti-nuclear movement and radical environmentalism become Luddite foolsplay smashing the job-machines of benevolent high technology; socialist programmes become anachron-istic visions of unobtainable utopias stupidly out-of-synch with an infi-nitely malleable capitalism. The end of modernism is joyfully proclaimed as if the creation of a radical and resistant postmodern political culture were impossible, as if the problems addressed by the various modern movements had disappeared, melted entirely into air.

The development of a radical political culture of postmodernism will accordingly require moving beyond rigorous empirical descriptions which imply scientific understanding but too often hide political meaning; beyond a simplistic anti-Marxism which rejects all the insights of historical materialism in the wake of an exposure of its contemporary weaknesses and gaps; beyond the disciplinary chauvinisms of an outdated academic division of labour desperately clinging to its old priorities; beyond a Marxist geography that assumes that a historical geographical materialism has already been created by merely inserting a second adjective. A new 'cognitive mapping' must be developed, a new way of seeing through the gratuitous veils of both reactionary postmodernism and late modern historicism to encourage the creation of a politicized spatial consciousness and a radical spatial praxis. The most important postmodern geographies are thus still to be produced.

# 3

# The Socio-spatial Dialectic

> Space and the political organization of space express social relationships but also react back upon them.... Industrialization, once the producer of urbanism, is now being produced by it.... When we use the words 'urban revolution' we designate the total ensemble of transformations which run throughout contemporary society and which bring about a change from a period in which questions of economic growth and industrialization predominate to the period in which the urban problematic becomes decisive.

These observations are drawn from a postscript to *Social Justice and the City* (1973, 306) in which David Harvey presented a brief appreciation and critique of the ideas of Henri Lefebvre on urbanism, the organization of space, and contemporary Marxist analysis. But Harvey's interpretation accomplished something more than a sympathetic introduction of Lefebvre to anglophonic Marxist geography. It recapitulated the pattern of response to Lefebvre's theory of space that had already appeared in French through Manuel Castells's important work, *La Question urbaine* (1972). Harvey praised Lefebvre but dissented from his insistence on the 'decisive' and 'pre-eminent' role of spatial structural forces in modern capitalist society. Both Harvey and Castells recognized Lefebvre's contribution in dealing brilliantly with the organization of space as a material product, with the relationship between social and spatial structures of urbanism, and with the ideological content of socially created space. But surely Lefebvre had gone too far? They both insinuated that he had elevated the urban spatial 'problematic' to an intolerably central and apparently autonomous position. The structure of spatial relations was being given an excessive emphasis while the more fundamental roles of production (versus circulation and consumption), social (versus spatial) relations of production, and industrial (versus finance) capital were being submerged within an overinterpreted

76

alternative – what Lefebvre called the 'urban revolution', *La Révolution urbaine* (1970). In his conceptualization of urbanism, Lefebvre appeared to them to be substituting spatial/territorial conflict for class conflict as the motivating force behind radical social transformation.

The key question to Harvey in 1973 was whether the organization of space (in the context of urbanism) was '*a separate structure* with its own laws of inner transformation and construction', or 'the *expression* of a set of relations embedded in some broader structure (such as the social relations of production)'. To Harvey – as to Castells previously – Lefebvre seemed to be a 'spatial separatist' and was thus succumbing to what might be called a fetishism of space. Struggling to be serious and rigorous in their application of Marxism, pioneers of Marxist geography like Harvey and Castells thus began to establish certain boundaries beyond which radical spatial analysis must not reach.

This pattern of response pervaded the new Marxist analysis of space that developed in the 1970s, significantly blunting its impact and weakening its accomplishments. The reaction to Lefebvre and the misunderstanding on his ideas was one manifestation of this rigidifying tendency. Indeed one can go a step further and argue that the first generation to develop a spatially explicit form of Marxist analysis – exemplified best in the pioneering works of Harvey and Castells but also in the rapidly expanding literature on radical urban and regional political economy (see Chapter 4) – was built upon an unnecessarily limited conceptualization of spatial relations. Thus what should have been the far-reaching implications of Marxist spatial analysis were unnecessarily blunted through the well-intended but short-sighted efforts of radical scholars to avoid the presumed dangers of spatial fetishism.

Rather ironically, the primary source of misunderstanding seemed to lie in the failure of Marxist analysts to appreciate the essentially dialectical character of social and spatial relationships as well as that of other structurally linked spheres like production and consumption. As a result, instead of sensitively probing the mix of opposition, unity, and contradiction which defines a socio-spatial dialectic, attention was too often drawn to empty categorical questions of causal primacy.[1] Within this

---

1. Typical of this impulse to protect the eternal primacy of the (nonspatial) social was Richard Walker's comment on my initial depiction of the socio-spatial dialectic in a paper on 'Topian Marxism', presented at the Annual Meetings of the Association of American Geographers in New Orleans, 1978. Walker, in an interesting paper on uneven development in advanced capitalism (1978), argued that dialectical analysis already incorporates the spatial relations of the mode of production, but that social relations (as value relations) remain primary. Value relations, however, are defined as abstract and aspatial — but nonetheless social. This depiction Walker himself described as 'undialectical and convenient' and I agree. It is precisely this 'convenient' suspension of dialectical reasoning that permits spatial relations to be incorporated yet immediately (and undialectically – or

rigidly categorical logic, it was difficult to see that the socio-spatial dialectic fitted neither of the two alternatives pressed upon Lefebvre by David Harvey. The structure of organized space is not a separate structure with its own autonomous laws of construction and transformation, nor is it simply an expression of the class structure emerging from social (and thus aspatial?) relations of production. It represents, instead, a dialectically defined component of the general relations of production, relations which are simultaneously social and spatial.

To establish this simultaneity, it must be clearly demonstrated that there exists a corresponding spatial homology to traditionally defined class relations and hence to the contingencies of class conflict and structural transformation. As I will attempt to demonstrate, such a space–to–class homology can be found in the regionalized division of organized space into dominant centres and subordinate peripheries, socially created and polarized spatial relations of production which are captured with greater precision in the concept of geographically uneven development. This conceptualization of the links between social and spatial differentiation does not imply that the spatial relations of production or the centre–periphery structure are separate and independent from the social relations of production, from class relations. On the contrary, the two sets of structured relations (the social and the spatial) are not only homologous, in that they arise from the same origins in the mode of production, but are also dialectically inseparable.

That there exists such a dialectical association between what might be called the vertical and horizontal dimensions of the mode of production, is suggested in the writings of Marx and Engels: in discussions of the antithesis between town and countryside, the territorial division of labour, the segmentation of urban residential space under industrial capitalism, the geographical unevenness of capitalist accumulation, the role of rent and private ownership of land, the sectoral transfer of surplus value, and the dialectics of nature. But one hundred years of Marxism have not been enough to develop the logic and scope of these insights.[2]

The atrophy of the geographical imagination in the intervening epoch helps explain why the rebirth of Marxist spatial analysis was so arduous and burdened with unfounded fears of spatial fetishism. The long hiatus also explains why there was so much controversy over terminology,

---

uncritically, if you will) subordinated to a despatialized notion of the social, apparently as a rigid structural universal evident in every historical moment in the development of capitalism.

2. One of the few attempts to explain why spatial analysis has historically been so little developed in Marxism can be found in Lefebvre's *La Pensée marxiste et la ville* (1972).

emphasis, and credentials; as well as why divisions persisted between urban, regional, and international political economies rather than leading to the creation of a more unified spatial political economy. Finally it helps us understand why, except for Lefebvre, there was such a lack of audacity – that is, why amid claims that the resurgence of spatially explicit, radical political economy represented a 'new' urban sociology, a 'new' economic geography, a 'new' urban politics, or a 'new' planning theory, no one else was ready to grasp the really radical implication that what was emerging was a dialectical materialism that is simultaneously historical and spatial. What follows is an attempt to recapture the initial assertion of the socio-spatial dialectic and the need for a historico-geographical materialism as it originally developed in Soja (1980) and Soja and Hadjimichalis (1979).

## Spatiality: the Organization of Space as a Social Product

It is necessary to begin by making as clear as possible the distinction between space *per se*, space as a contextual given, and socially-based spatiality, the created space of social organization and production. From a materialist perspective, whether mechanistic or dialectical, time and space in the general or abstract sense represent the objective form of matter. Time, space, and matter are inextricably connected, with the nature of this relationship being a central theme in the history and philosophy of science. This essentially physical view of space has deeply influenced all forms of spatial analysis, whether philosophical, theoretical or empirical, whether applied to the movement of heavenly bodies or to the history and landscape of human society. It has also tended to imbue all things spatial with a lingering sense of primordiality and physical composition, an aura of objectivity, inevitability, and reification.

Space in this generalized and abstracted physical form has been conceptually incorporated into the materialist analysis of history and society in such a way as to interfere with the interpretation of human spatial organization as a social product, the key first step in recognizing a socio-spatial dialectic. Space as a physical context has generated broad philosophical interest and lengthy discussions of its absolute and relative properties (a long debate which goes back to Leibniz and beyond), its characteristics as environmental 'container' of human life, its objectifiable geometry, and its phenomenological essences. But this physical space has been a misleading epistemological foundation upon which to analyse the concrete and subjective meaning of human spatiality. Space in itself may be primordially given, but the organization, and meaning of space is a product of social translation, transformation, and experience.[3]

Socially-produced space is a created structure comparable to other social constructions resulting from the transformation of given conditions inherent to being alive, in much the same way that human history represents a social transformation of time. Along similar lines, Lefebvre distinguishes between Nature as naively given context and what can be termed 'second nature', the transformed and socially concretized spatiality arising from the application of purposeful human labour. It is this second nature that becomes the geographical subject and object of historical materialist analysis, of a materialist interpretation of spatiality.

> Space is not a scientific object removed from ideology and politics; it has always been political and strategic. If space has an air of neutrality and indifference with regard to its contents and thus seems to be 'purely' formal, the epitome of rational abstraction, it is precisely because it has been occupied and used, and has already been the focus of past processes whose traces are not always evident on the landscape. Space has been shaped and molded from historical and natural elements, but this has been a political process. Space is political and ideological. It is a product literally filled with ideologies. (1976b, 31)

## Organized space and the mode of production: three points of view

Once it becomes accepted that the organization of space is a social product – that it arises from purposeful social practice – then there is no longer a question of its being a separate structure with rules of construction and transformation that are independent from the wider social framework. From a materialist perspective, what becomes important is the relationship between created, organized space and other structures within a given mode of production. It is this basic issue that divided Marxist spatial analysis in the 1970s into at least three distinctive orientations.

First, there were those whose interpretations of the role of organized

---

3. The dominance of a physicalist view of space has so permeated the analysis of human spatiality that it tends to distort our vocabulary. Thus, while such adjectives as 'social', 'political', 'economic', and even 'historical' generally suggest, unless otherwise specified, a link to human action and motivation, the term 'spatial' typically evokes a physical or geometrical image, something external to the social context and to social action, a part of the 'environment', a part of the setting for society – its naively given container – rather than a formative structure created by society. We really do not have a widely used and accepted expression in English to convey the inherently social quality of organized space, especially since the terms 'social space' and 'human geography' have become so murky with multiple and often incompatible meanings. For these and other reasons, I have chosen to use the term 'spatiality' to specify this socially-produced space.

space led them to challenge prevailing Marxist approaches, especially with regard to definitions of the economic base and superstructure. Again, Lefebvre offered a key argument:

> Can the realities of urbanism be defined as something superstructural, on the surface of the economic base, whether capitalist or socialist? No. The reality of urbanism modifies the relations of production without being sufficient to transform them. Urbanism becomes a force in production, rather like science. *Space and the political organization of space express social relationships but also react back upon them.*[4]

Here we have opened the possibility of a complex socio-spatial dialectic operating within the structure of the economic base, in contrast with the prevailing materialist formulation which regards the organization of spatial relations only as a cultural expression confined to the superstructural realm. The key notion introduced by Lefebvre in the last sentence becomes the fundamental premise of the socio-spatial dialectic: that social and spatial relations are dialectically inter-reactive, interdependent; that social relations of production are both space-forming and space-contingent (at least insofar as we maintain, to begin with, a view of organized space as socially constructed).

Within a regional as opposed to urban frame, similar ideas were developed by Ernest Mandel. In his examination of regional inequalities under capitalism, Mandel (1976, 43) declared that 'The unequal development between regions and nations is the very essence of capitalism, on the same level as the exploitation of labour by capital'. By not subordinating the spatial structure of uneven development to social class but viewing it as 'on the same level', Mandel identified for the regional and international scale a spatial problematic that closely resembled Lefebvre's interpretation of urban spatiality, even to the point of suggesting a powerful revolutionary force arising from the spatial inequalities which he clearly claimed were necessary for capitalist accumulation. In his major work, *Late Capitalism* (1975), Mandel focused upon the crucial historical importance of geographically uneven development in the accumulation process and thereby in the survival

---

4. This observation, with added emphasis, comes from Harvey's (1973) translation of a segment of *La Révolution urbaine* (1970, 25). At this point in the development of his ideas on the production of space, Lefebvre had fastened on to urbanism as a summative conceptualization of capitalist spatiality. Unfortunately, this explicitly urban metaphor prevented readers from seeing the much more general spatial emphasis that lay behind his developing argument and provoked responses to a perceived reification of the urban. Castells would crystallize this view by describing Lefebvre's conceptualization of the urban revolution as the left-wing version of the 'urban ideology' promulgated by the bourgeois theoreticians of the Chicago School of urban ecology, which he considered an equally mystifying overspecification of the urban as theoretical object.

and reproduction of capitalism itself. In so doing, he presented one of the most rigorous and systematic Marxist analyses of the political economy of regional and international development ever written.

Neither Lefebvre nor Mandel, however, fully succeeded in defining a cross-scalar synthesis of the socio-spatial dialectic and their formulations thus remained incomplete. Nevertheless, in their attribution to the structure of spatial relations of a significant transformational potential in capitalist society comparable to that which has conventionally been associated with the 'vertical' class struggle, the direct social conflict between labour and capital, both Lefebvre and Mandel presented a point of view which provoked strong resistance from other Marxists who saw the spectre of spatial determinism rising again.

This resistance to the suggestion that organized space represents anything more than a reflection of the social relations of production, that it can engender major contradictions and transformational potential with regard to the mode of production, that it is some way homologous to class structure and relations, defined another, much larger group of radical scholars. Included here was the growing cadre of critics seeking to maintain some form of Marxist orthodoxy by persistent screening of the 'new' urban and regional political economy. Characteristic to this group was the belief that neo-Marxist analysis adds little that is inherently new to more conventional Marxist approaches, that the centrality of traditional class analysis is inviolable, and thus that neo-Marxist urban and regional analyses, while interesting, are too often unacceptably revisionist and analytically muddled. Needless to say, the conceptualization (or non-conceptualization) of space adhered to by this group deviated little from the traditional historicism of Marxism after Marx.

A third approach can be identified, however, falling somewhere in between these two extremes. Its practitioners appeared, implicitly at least, to be adopting much the same formulation as described by Lefebvre and Mandel. Yet when pushed to an explicit stance, they maintained the pre-eminence of aspatial social class definitions, sometimes to the point of tortuously trying to resist the implications of their own analyses. In this group were Manuel Castells, David Harvey, Immanuel Wallerstein, André Gunder Frank, and Samir Amin, all of whom contributed insightful depictions of the socio-spatial dialectic as I have defined it. Each, however, backed off from an open recognition of the formative significance of spatiality into analytically weak and vulnerable positions on the role of spatial structure in the development and survival of capitalism. Whereas the first group mentioned occasionally overstated the socio-spatial dialectic, this group retreated from it without effectively capturing its meaning and implications, creating a confusing ambivalence – which was reacted to in turn by the more orthodox Marxist critics.

To take a prominent example, consider Castells's conceptualization of space in *The Urban Question* (1977), a book purposefully titled to contrast with *The Urban Revolution*, written by his former teacher, Lefebvre.

> To consider the city as the projection of society on space is both an indispensable starting point and too elementary an approach. For, although one must go beyond the empiricism of geographical description, one runs the very great risk of imagining space as a white page on which the actions of groups and institutions are inscribed, without encountering any other obstacle than the trace of past generations. This is tantamount to conceiving of nature as entirely fashioned by culture, whereas the whole social problematic is born by the *indissoluble union* of these two terms, through the *dialectical process* by which a particular biological species (particularly because divided into classes), "man", transforms himself and transforms his environment in his struggle for life and for the differential appropriation of the product of his labour.
>
> *Space is a material product*, in relation with other elements – among others, men, who themselves enter into particular social relations, which give to space (and to the other elements of the combination) a form, a function, a social signification. It is not, therefore, a mere occasion for the deployment of social structure, but a concrete expression of each historical ensemble in which a society is specified. It is a question, then, of establishing, in the same way as for any other real object, the *structural and conjunctural laws that govern its existence and transformation*, and the specificity of its articulation with the other elements of a historical reality. This means that there is no theory of space that is not an integral part of a general social theory, even an implicit one. (115, emphasis added)

This complex passage encompasses a socio-spatial dialectic but is presented as an alternative to the rejected Lefebvrean view. No wonder the readers of the English translation were confused. Castells's conceptualization itself was attacked as revisionist and Weberian by representatives of the second camp. Harloe (1976, 21), for example, claimed that Castells committed the same error that he criticized Lefebvre for falling into, that is, separating the spatial structure from its roots in production and class relations. This presumed error, it was argued, engendered an inappropriate emphasis on collective consumption and other social and spatial aspects of the consumption process, an emphasis which was seen as confounding the more fundamental role of production in capitalist urbanization.[5]

---

5. Interestingly enough, a very similar reaction was developing at about the same time to Wallerstein and others who were attempting to give an explicit spatial dimension to the international division of labour and the uneven development of the world capitalist

But let us return to Castells's principal contribution to what he called 'the debate on the theory of space'. Castells clearly presented space as a material product emerging dialectically from the interaction of culture and nature. Space was thus not simply a reflection, a 'mere occasion for the deployment', of the social structure, but the concrete expression of a combination of instances, an 'historical ensemble' of interacting material elements and influences. How then can one understand and interpret this created space? The route was through what Castells described as the 'structural and conjunctural laws that govern its existence and trans-formation', a hard indication of the Althusserian structuralism that was then governing Castells's approach to the urban question.

What appeared to separate Castells and Lefebvre was the former's unqualified argument that 'particular social relations' give form, func-tion, and significance to the spatial structure and all other 'elements of the combination'. One 'structure' – the supposedly aspatial social relations of production (which somehow include property rights while ignoring their spatial/territorial dimension) – was thus given a deter-minant role. But it is precisely this determinative relationship which Lefebvre began to qualify and amend by associating class formation with both social and spatial relations of production and embedding the 'social problematic' in a simultaneously social and spatial division of labour, a vertical and horizontal dimension. There did not yet exist in the 1970s a rigorous formulation of these spatial relations of production and spatial divisions of labour, certainly not one which matched the depth and persuasiveness of Marxist analyses of the social relations of production and the social divisions of labour. But there was no reason to reject the formulation of a socio-spatial dialectic on the grounds that a century of Marxism failed to incorporate a materialist interpretation of spatiality to match its materialist interpretation of history.

### The origins of the neglect of spatiality in Western Marxism

It became common practice amongst Marxist geographers and urban sociologists in the 1970s to argue that within the classic works of Marx, Engels, and Lenin, there are powerful geographical and spatial intuitions

---

economy. They too were attacked for their overemphasis on consumption and exchange (versus production), their unrecognized reversion to bourgeois mystifications of class (via Adam Smith rather than Max Weber), their inappropriate spatializations of history and capitalist development (that is, their excessive emphasis on forces external to the evolving social relations of production *in situ* via the core–periphery structure and the global-scale operations of capitalist accumulation). I will discuss this critical debate in the next chapter.

but that these emphases and orientations remained weakly developed by successive generations. Many thus approached the task of Marxist spatial analysis largely in terms of drawing out and elaborating upon these classical observations in the context of contemporary capitalism. David Harvey's analysis of the geography of capitalist accumulation (1975) and Jim Blaut's work on imperialism and nationalism (1975) are excellent examples, while larger projects aimed at extracting the geographical implications of Marx's writings were established under the direction of the contributors to *Antipode* and members of its linked organization, the Union of Socialist Geographers.

Relatively little attention, however, was given to explaining why spatial analysis remained so weakly developed for so long a time. Indeed, until recently, Western Marxism paralleled the development of bourgeois social science in viewing the organization of space as either a 'container' or an external reflection, a mirror of the social dynamic and social consciousness. In an almost Durkheimian manner, the spatiality of social life was externalized and neutered in terms of its impact on social and historical processes and seen as little more than backdrop or stage. Explaining this submergence of spatial analysis in Marxism remains a major task. It is possible, however, to establish some initial theses.

1. *The late appearance of Grundrisse.* Marx's *Grundrisse*, which was not disseminated widely in translation until well after the Second World War, probably contains more explicit geographical analysis than any of his writings. Its two volumes were first published in Russian in 1939 and 1941. The first German edition appeared in 1953, the first English edition in 1973. In addition, as is now well known, Marx never completed his plans for subsequent volumes of *Capital* dealing with world trade and the geographical expansion of capitalism, only hinting at their possible content in the late-appearing *Grundrisse.* In the absence of these sources, heavy emphasis was placed upon the largely aspatial, closed-system theorizing of the published volumes of *Capital.* Although Marx never fails to illustrate his arguments with specific historical and geographical examples, volumes I and II of *Capital* in particular remain encased in the simplifying assumptions of a closed national economy, an essentially spaceless capitalism systematically structured almost as if it existed on the head of a pin. Volume III and the proposed additional volumes were to represent concretizations of Marx's theory, projections outward into the historical and geographical analysis of world markets, colonialism, international trade, the role of the state, etc. – in essence, toward an analysis of the uneven development of productive sectors, regions, and nations.

Through the contributions of Bukharin, Lenin, Luxemburg, Trotsky and others, the theory of imperialism and associated conceptualizations of the processes of uneven development became the major context for geographical analysis within Western Marxism. There was an implied spatial problematic in these theorizations of imperialism, but it rested primarily on the simple recognition of an ultimate physical limitation to the geographical expansion of capitalism. For most of the major theoreticians, these geographical limits to capital were unlikely ever to be reached because social revolution would intervene long before the entire world became uniformly capitalist. Nevertheless, the processes of geographically uneven development were recognized and placed on the theoretical and political agenda, to be revived by a new generation of Marxist scholarship, led by such figures as Wallerstein, Amin, Emmanuel, Palloix, Hymer, and, especially, Ernest Mandel. How much this new generation was influenced by the post-war translations of *Grundrisse* remains an interesting but still open question.

2. *Anti-spatial traditions in Western Marxism.* Failure to develop the spatial emphases inherent in Marx's and later works on the geographical expansion of capitalism and in the equally spatial interpretations of the town-countryside antithesis that appear so vividly in *The German Ideology* and elsewhere in Marx's writings can also be linked to a deep tradition of anti-spatialism. Paradoxically perhaps, this tradition of rejecting geographical explanations of history originates with Marx himself, in his response to the Hegelian dialectic.

In many ways, Hegel and Hegelianism promulgated a powerful spatialist ontology and phenomenology, one which reified and fetishized space in the form of the territorial state, the locus and medium of perfected reason. As Lefebvre argues in *La Production de l'espace* (1974, 29–33), for Hegel historical time became frozen and fixed within the imminent rationality of space as state-idea. Time thus became subordinated to space, with history itself being directed  by a territorial 'spirit', the state. Marx's anti-Hegelianism was not confined to a materialist critique of idealism. It was also an attempt to restore historicity – revolutionary temporality – to primacy over the spirit of spatiality. From this project arose a powerful sensitivity and resistance to the assertion of space into a position of historical and social determination, an anti-Hegelian anti-spatialism which is woven into virtually all of Marx's writings.

The possibility of a 'negation of the negation', a non-prioritized recombination of history and geography, time and space, was buried under subsequent codifications of Marx's theory of fetishism. A historical materialist dialectic was embraced as human beings were contextualized in the making of history; but a spatial dialectic, even a materialist one, with human beings making their geographies and being constrained by what they have made, was unacceptable. This form of anti-spatialism may have been most rigidly codified by Lukacs in *History and Class Consciousness*, wherein spatial consciousness is presented as the epitome of reification, as false consciousness manipulated by the state and by capital to divert attention away from class struggle.

This anti-spatial armour served well to resist the many attacks on Marxism and the working class based on uncontestable spatial reification – Le Corbusier's choice between 'Architecture or Revolution' being among the most innocuous of these attacks, fascism by far the most vicious – but it also tended to associate all forms of spatial analysis and geographical explanation with fetishism and false consciousness. Not only does this tradition continue to interfere with the development of Marxist spatial analysis, it has also been partially responsible for the characteristic confusion surrounding the formulation of a sufficiently concretized Marxist theory of the state, nationalism, and local politics.

Mention must also be made of the anti-spatial character of Marxist dogmatism as it emerged from the Second International and was consolidated under Stalinism. Spatial questions, among many other aspects of Marxist theory and practice, were treated by the Second International and its leaders through the scope of a sterile economic reductionism. Marxism was turned into positivistic scientism under Stalin, emphasizing a belief in technocratic thought and strictly economic causality in the links between base and superstructure. Culture, politics, consciousness, ideology, and, along with them, the production of space were reduced to simple reflections of the economic base. Spatiality became absorbed in economism as its dialectical relationship with other elements of material existence was broken.

3. *Changing conditions of capitalist exploitation.* The early neglect and recent revival of interest concerning the spatial problematic in Marxism may, in the end, be primarily a reflection of changing material conditions. Lefebvre has argued, in *La Pensée marxiste et la ville* (1972), that during the nineteenth century and into the early twentieth the spatial problematic was simply less important than it is today with respect to both the exploitation of labour and the reproduction of the essential means of production. Under the conditions of competitive industrial capitalism, machines, commodities, and the labour force were reproduced under specific social legislation (labour contracts, civil laws, technological agreements) and an oppressive state apparatus (police, the military, colonial administration). The production of space was accommodative, conformal, and directly shaped by the market and state power. The spatial structure of the industrial capitalist city, for example,

repeated itself over and over again in its functional concentricity and segregated sectors of social class.

Exploitation and social reproduction were primarily embedded in a manipulable matrix of time. The rate of exploitation, Marx's ratio between surplus value and variable capital, is, after all, an expression derived from the labour theory of value and its fundamental measurement of socially necessary labour time. Like the formulas for the organic composition of capital and the rate of profit, its derivation assumes a closed system view of capitalist production relations, devoid of significant spatial differentiation and unevenness. In addition, given the massive urbanization associated with expanding industrialization, the reproduction of the labour force was much less crucial an issue than the process of direct exploitation through a system of subsistence wages and the domination of capital over labour at the point of production. In the extraction of absolute surplus value, the social organization of time appeared to be more important than the social organization of space.

In contemporary capitalism (setting aside for the moment the question of transition and restructuring, its causes, timing, and so on) the conditions which underlie the continuing survival of capitalism have changed. Exploitation of labour time continues to be the primary source of absolute surplus value, but within increasing limits arising from reduction in the length of the working day, minimum wage levels and wage agreements, and other achievements of working-class organization and urban social movements. Capitalism has been forced to shift greater and greater emphasis to the extraction of relative surplus value through technological change, modifications in the organic composition of capital, the increasingly pervasive role of the state, and the net transfers of surplus associated with the penetration of capital into not fully capitalist spheres of production (internally, through intensification, as well as externally, through uneven development and geographical 'extensification' into less industrialized regions around the world). This has required the construction of total systems to secure and regulate the smooth reproduction of the social relations of production. In this process, the production of space plays a crucial role. It is this switch in significance between the temporality and spatiality of capitalism that provoked Lefebvre to argue that 'industrialization, once the producer of urbanism, is now being produced by it'.

## Defining the Spatial Problematic

The development of systematic Marxist spatial analysis coincided for the most part with the intensification of social and spatial contradictions in both core and peripheral countries due to the general crisis in capitalism beginning in the 1960s. But well before this period, there were several important precursors within the Western Marxist tradition that should not be overlooked. Most commonly, the theories of imperialism are seen as the primary source of spatial thinking in Western Marxism. There were, however, several other significant progenitors.

For example, in the USSR between 1917 and 1925, an *avant garde* movement of city planners, geographers, and architects worked toward achieving a 'new socialist spatial organization' to correspond with other revolutionary movements in Soviet society (see Kopp, 1971). Spatial transformation was not assumed to be an automatic byproduct of revolutionary social change. It too involved struggle and the formation of a collective consciousness. Without such effort, the prerevolutionary organization of space would continue to reproduce social inequality and exploitational structures. The innovative activities of this group of radical spatial thinkers were never fully accepted, and eventually their revolutionary experiment in the socialist reconstruction of space was abandoned in the drive towards industrialization and military security under Stalin. Productivism and military strategy came to dominate spatial policy in the Soviet Union, all but burying the significance of a more profound spatial problematic in socialist transformation.

## The precursory contributions of Antonio Gramsci

Another important but often neglected contribution to the development of Marxist spatial analysis can be found in the work of Antonio Gramsci. In part, Gramsci's work relates to the contemporary situation because it contains some well elaborated analyses of urban and regional problems in Europe during the 1920s and the early stages of the Great Depression. But of even greater relevance than these specific studies of regional backwardness in the Mezzogiorno, urban development in Turin, the housing question, and the evolving alliances between the urban and rural proletariat, was a more general effort to focus attention upon the political, cultural, and ideological dimensions of capitalism (against the prevailing economism of the time) and, especially, to elaborate upon the role of the modern capitalist state and its imposed territorial division of labour.

Gramsci, in his emphasis on the 'ensemble of relations' which comprise a particular social formation, concretized the mode of production in time and space, in history and geography, in a specified conjunctural framework which became the necessary context for revolutionary strategy. A spatial problematic was not explicitly raised as such, but its foundations were clearly evident in the spatial relations embedded in the

social formation and in the particularities of place, location, and territorial community.

Revolutionary strategy for Gramsci is situated in three interconnected arenas, all linked in one way or another to the spatiality of
social life under capitalism. First, in his analyses of the political and
ideological structures of the Italian social formation, one can find the
seeds of contemporary theorizations of the capitalist state and its dual
and contradictory functions of repression/legitimization and material/
ideological reproduction. His stress on hegemony and his work on popular culture, state control over everyday life, the importance of local
'council' organizations, and the relation between occupational and territorial structures reflects an implicit understanding of the socio-spatial
dialectic.

The first area of emphasis relates to the second: the role of exploitation of the working class at their place of residence, the point of
consumption and reproduction versus the point of production, the
workplace. Gramsci's writings not only reopened the 'housing question'
to fresh consideration but directly challenged both the economism and
productivism of the Second International and the 'workerism' of the
Italian Socialist and Communist parties of the time. It also prefigured
the rise of a new urban and regional political economy focused on local
struggles over collective consumption and the mobilization of urban,
rural, and regional social movements.

Finally, these two strategic emphases came together again in
Gramsci's conceptualization of the revolutionary historical bloc, an
alliance of popular movements fighting for similar goals and linked
conjuncturally to the specific conditions of capitalist crisis. These conditions were not only economic, but political, cultural, and ideological as
well; they combined both production and reproduction, the workplace
and the residential community. In *Prison Notebooks*, Gramsci saw the
growing complexity of modern capitalist society and the need to raise
political, cultural, and ideological struggles to a new level as the state
seemed increasingly to rely on its legitimizing hegemony rather than on
direct force and oppression. Revolutionary consciousness thus came to
be rooted in 'the phenomenology of everyday life'.

The step from Gramsci to Lefebvre is primarily one of explicitness
and emphasis regarding the spatialization of this phenomenology of
everyday life. Lefebvre, like Gramsci, fought consistently against
dogmatic and reductionist interpretations of Marxism and reasserted the
multifaceted exploitation embedded in *la vie quotidienne* as the basis for
a critique of the *ouvrièrisme* of the modern left: the narrow emphasis on
exploitation and struggle at the workplace and hence on such totalizing
strategies as the general strike. For Lefebvre, echoing Gramsci, 'the

revolution can only take place conjuncturally, i.e., in certain class relations, an ensemble of relations into which the peasantry and the intellectual enter'. (1976a, 95) Lefebvre, however, moves on to 'spatialize the conjuncture' and thus inserts a spatial problematic into the centre of revolutionary consciousness and struggle.

## The spatial problematic and the survival of capitalism

Lefebvre's writings are marked by a persistent search for a political understanding of how and why capitalism has survived from the competitive industrial form of Marx's time to the advanced, state-managed and oligopolistic industrial capitalism of today. As described in Chapter 2, he presented a series of increasingly elaborated 'approximations' beginning with his conceptualization of everyday life in the modern world and passing through revolutionary urbanization and urbanism to his major thesis on the social production of space. This thesis is pointedly condensed in *The Survival of Capitalism* (1976a, 21), the only one of Lefebvre's explicitly spatialized texts to have been translated into English.

> Capitalism has found itself able to attenuate (if not resolve) its internal contradictions for a century, and consequently, in the hundred years since the writing of *Capital*, it has succeeded in achieving "growth." We cannot calculate at what price, but we do know the means: *by occupying space, by producing a space.*

Lefebvre links this advanced capitalist space directly to the reproduction of the social relations of production, that is, the processes whereby the capitalist system as a whole is able to extend its existence by maintaining its defining structures. He defines three levels of reproduction and argues that the ability of capital to intervene directly and affect all three levels has developed over time, with the development of the productive forces. First, there is bio-physiological reproduction, essentially within the context of family and kinship relations; second is the reproduction of labour power (the working class) and the means of production; and third is the still broader reproduction of the social relations of production. Under advanced capitalism the organization of space becomes predominantly related to the reproduction of the dominant system of social relations. Simultaneously, the reproduction of these dominant social relations becomes the primary basis for the survival of capitalism itself.

Lefebvre grounds his argument in the assertion that socially produced space (essentially urbanized space in advanced capitalism, even in the countryside) is where the dominant relations of production are reproduced. They are reproduced in a concretized and created spatiality that has been progressively 'occupied' by an advancing capitalism, fragmented into parcels, homogenized into discrete commodities, organized into locations of control, and extended to the global scale. The survival of capitalism has depended upon this distinctive production and occupation of a fragmented, homogenized, and hierarchically structured space – achieved largely through bureaucratically (that is to say, state) controlled collective consumption, the differentiation of centres and peripheries at multiple scales, and the penetration of state power into everyday life. The final crisis of capitalism can only come when the relations of production can no longer be reproduced, not simply when production itself is stopped (the abiding strategy of *ouvrièrisme*).

Thus, class struggle (yes, it still remains class struggle) must encompass and focus upon the vulnerable point: the production of space, the territorial structure of exploitation and domination, the spatially controlled reproduction of the system as a whole. And it must include all those who are exploited, dominated, and 'peripheralized' by the imposed spatial organization of advanced capitalism: landless peasants, proletarianized petty bourgeoisies, women, students, racial minorities, as well as the working class itself. In advanced capitalist countries, Lefebvre argues, the struggle will take the form of an 'urban revolution' fighting for *le droit à la ville* and control over *la vie quotidienne* within the territorial framework of the capitalist state. In less industrialized countries, it will also focus upon territorial liberation and reconstruction, on taking control over the production of space and its polarized system of dominant cores and dependent peripheries within the global structure of capitalism.

With this chain of arguments, Lefebvre defines an encompassing spatial problematic in capitalism and raises it to a central position within class struggle by embedding class relations in the configurative contradictions of socially organized space. He does not argue that the spatial problematic has always been so central. Nor does he present the struggle over space as a substitute for, or alternative to, class struggle. Instead, he argues that no social revolution can succeed without being at the same time a consciously spatial revolution. In much the same way as other 'concrete abstractions' (such as the commodity form) have been analysed in the Marxist tradition to show how they contain within them, mystified and fetishized, the real social relations of capitalism, so too must we now approach the analysis of space. The demystification of spatiality will reveal the potentialities of a revolutionary spatial consciousness, the material and theoretical foundations of a radical spatial praxis aimed at expropriating control over the production of space. Berger's claim comes back again: 'prophesy now involves a geographical rather than historical projection; it is space, not time, that hides consequences from us'.

# 4

# Urban and Regional Debates: the First Round

> The entire capitalist system ... appears as a hierarchical structure of different levels of productivity, and as the outcome of the uneven and combined development of states, regions, branches of industry and firms, unleashed by the quest for surplus profit.... Thus even in the 'ideal case' of a homogeneous beginning, capitalist economic growth, extended reproduction and accumulation of capital are still synonymous with the juxtaposition and constant combination of development and underdevelopment. (Mandel, 1975, 102 and 85)

## The Urban Spatial Problematic

Marxist spatial analysis at the urban scale evolved through the 1970s in conjunction with a larger development that combined several disciplinary emphases (economic, sociological, geographical) into a common focus on the political economy of urbanization. Underlying this integrative development was a set of assumptions about the changing nature of the urbanization process in advanced capitalism. The rising importance of a monopoly capital, its expansion on a global scale, and its increasing dependence on state management and planning were interpreted as having introduced new historical (and spatial) conditions into contemporary capitalist social formations and hence into the politics of class struggle. Among other effects, these new conditions demanded a different approach to the city and to the urbanization process than that which had characterized the treatment of urban problems under the competitive capitalism of Marx's time. The urbanization process became a revealing social hieroglyphic through which to unravel the dynamics of post–war capitalist development and to strategize an appropriate political response to an increasingly urbanized world economy.

94

The city came to be seen not only in its distinctive role as a centre for industrial production and accumulation, but also as the control point for the reproduction of capitalist society in terms of labour power, exchange, and consumption patterns. Urban planning was critically examined as a tool of the state, serving the dominant classes by organizing and reorganizing urban space for the benefit of capital accumulation and crisis management. Major attention was given not only to contradictions at the place of work (the point of production) but also to class conflict over housing and the built environment, the state provisioning and siting of public services, community and neighbourhood economic development, the activities of financial organizations, and other issues which revolved around how urban space was socially organized for consumption and reproduction. A specifically urban spatial problematic – incorporated into the dynamics of urban social movements – was thus put on the agenda for both theoretical consideration and radical social action.

More orthodox Marxists discerned a disturbing and potentially destructive revisionism, especially with regard to the traditional primacy of production issues. In many ways, this was an understandable reaction. Efforts to separate consumption from production and to assign to it a significant autonomous strength in society and history – to define class, for example, primarily upon consumption characteristics – typified bourgeois or liberal social science and its response to historical materialism. The presence of a strong neo-Weberian current in the new urban political economy made Marxists even more wary of its intent and focus. Furthermore, the emphasis given to spatial analysis triggered fears of a recrudescence of geographical determinism. Insofar as the new urban political economy occasionally did fall into a 'consumptionist' tangent and disconnected spatial relations from their origin in the relations of production, it deserved the forceful critical response.

For the most part, however, the growing emphasis on distribution, exchange, and collective consumption did not so much represent a denial of the central role of production as a call for greater attention to certain processes that historically had been neglected in Marxist analysis. Today, it was argued, these processes have become more pertinent to class analysis and class conflict and deserved more direct and intensive consideration. In this sense, Marxist urban analysis was solidly within the critical tradition of Western Marxism that had evolved through the twentieth century against the grain of formalistic, infrastructural economism (see Anderson, 1976).

Moreover, it could be (and was) argued that Marx himself viewed production and consumption as dialectically related moments of the same process, each necessarily requiring the other. Surplus-value, arising out of the application of human labour to the production of commodities, was the essential substance of capitalist accumulation. But such surplus-value remained an abstract potential unless and until it was realized through the nexus of exchange and thus through the consumption process. To break this linked chain and to erect one moment timelessly above the others would be to oversimplify Marx's own conceptualization of the process (which was most explicitly described in *Grundrisse*, 1973, 90–94).

But whereas the emphasis on consumption and reproduction became effectively legitimized, much less successful was the attempt to defend the integrative spatial perspective that had entered Marxist urban analysis. As described in the previous chapter, the old fear of fetishizing space hamstrung efforts to introduce spatiality more centrally within the larger context of dialectical and historical materialism and into the historical debates on the development and survival of capitalism. As a result, Marxist spatial analysis prospered, but primarily as an adjunct, a methodological emphasis, to what became the dominant focus of urban political economy, the search for a Marxist theory of the advanced capitalist state. Only Lefebvre's work penetrated the confusion and ambivalence to project, without compromise, a forceful urban spatial problematic into the core of contemporary Marxism.

The key to the Lefebvrean assertion was his recognition of a profound evolutionary transformation linked to the survival of capitalism into the twentieth century. This is what he refers to when claiming that we are in a period in which the urban problematic has become more politically decisive than questions of industrialization and economic growth. In contrast to an earlier time, when industrialization produced urbanism, we are now faced with a situation in which industrialization and economic growth, the foundations of capitalist accumulation, are shaped primarily by and through the social production of urbanized space, planned and orchestrated with increasing power by the state, and expanding to encompass more and more of the world's population and resources. The urban social movements that were receiving such contemporary attention were essentially rooted in the political response of those subordinated, peripheralized, and exploited by the particularities of this increasingly global spatial planning process.

Once this assertive stance is understood and accepted, it becomes possible to clarify the meaning and intent of several more specific arguments presented by Lefebvre. Take, for example, his comments on the historical relation between the primary (industrial) and secondary (financial) circuits in the circulation of surplus-value, as translated by David Harvey in *Social Justice and the City* (1973, footnote 1, 312):

Whereas the proportion of global surplus value formed and realized in industry declines, the proportion realized in speculation and in construction and
real estate grows. The secondary circuit comes to supplant the principal
circuit.

The secondary circuit is deeply involved in the manipulation of the built
environment, the extraction of urban rent, the setting of land values, and
the organization of urban space for collective consumption, in all cases
facilitated by the local and national state. Lefebvre does not claim that
surplus-value is created in this secondary circuit, only that the proportion realized therein has massively expanded to reflect the growing need
for direct intervention in the production of urban space.

These arguments suggest that the proportion of surplus-value that is
absorbed in the reproduction of labour power and the social relations of
production in an increasingly urbanized, increasingly monopolistic,
increasingly global, capitalist society has become larger than ever before
– perhaps even larger than that which is directly realized (and re-
invested) from industrial production, from the exploitation of labour by
capital at the workplace. This does not necessarily mean that the
absolute volume of surplus-value produced in industry has declined,
although the diversion of capital into non-productive activities is a significant structural problem in what Lefebvre described as the *système
étatique*, and others termed 'state monopoly capitalism'. What is argued
instead is that the increasing surplus product yielded through the
centralization and accumulation of capital on a global scale has dispro-
portionately increased the costs of reproducing labour and maintaining
capitalist social relations. The spatial planning of capitalist accumulation
and growth, the ability of capital to 'attenuate' (if not resolve) its internal contradictions' for the past century '*by occupying space, by producing a space*', has not been a free ride. Spatialization is expensive.
Taking the argument one step further, Lefebvre (and others, like
Castells) shifted attention to the urban arena of collective consumption
as a major site both for the realization of value and for an increasingly
spatialized class struggle.

A particularly clear description of the historical change in class
relations associated with the rise of state-managed monopoly capitalism
was presented by Shoukry Roweis, an urban planner based at the
University of Toronto:

We find a pronounced shift in the locus of class conflicts accompanying the
shift from early to late capitalism. The main problems in early capitalism were
problems of production (i.e. problems of insufficient aggregate supply, hence
relative scarcity). Under such conditions, class conflicts were over the division
of surplus production. Concretely, the conflicts were strictly labour/capital

conflicts centered in the workplace and focussing on wage/profit disputes.

In today's capitalism, this is no longer strictly true. *Problems of production gave way to problems of overproduction.... What labour takes away with one hand (in the workplace struggle) it gives away with the other (in the urban living place)....* With the intensified and expanded state intervention, the struggles over wages lose their meaning, and the struggle over political/ administrative power begins to impose itself as crucial, but yet unpursued struggle. At the same time, with the ever increasing urbanization of the population, most of these struggles acquire a definite urban character. *The struggle, in brief, has shifted from the sphere of production (of commodities and services) to the sphere of reproduction (i.e. the maintenance of stable, if not improving, standards of urban living).* (Roweis, 1975, 31–2, emphasis in original)

Competitive industrial capitalism has been able to extend and transform itself through a series of structural changes, but it has done so without eliminating its fundamental contradictions. They have instead become expressed not only in the direct confrontation of capital and labour at the point of production but also, and as intensively, in the realm of collective consumption and social reproduction. Whereas, under competitive capitalism, Engels could argue that consumption-based working-class militancy, as in the conflict over the 'housing question', could be resolved by the bourgeoisie to its own benefit (either by playing off improved housing conditions and reduced rents against increased wages or by simply shifting bad housing conditions to another location rather than eliminating them – the tactic of spatial 'displacement'), such consumption-based conflict can no longer be co-opted so easily. More than in the past, the realization of surplus-value, and hence the accumulation of capital itself, has become as dependent upon control over the means of consumption/reproduction as upon control over the means of production, even if this control rests ultimately in the same hands.

It can thus be argued that the potential transformation of capitalism has come increasingly to revolve around a simultaneously social and spatial struggle, a combined wage and consumption-based conflict, an organization and consciousness of labour as both workers and consumers, in other words a struggle arising from the exploitative structures inherent in both the vertical and horizontal class divisions of society, in a socio-spatial dialectic. Resistance and struggle in the contemporary context thus involve the articulation of social and spatial praxis. The categorical prioritization of the first over the second term in each of these combinations can no longer be accepted.

To illustrate more specifically the preceding arguments and to exemplify again the tripartite typology of approaches to Marxist spatial analysis that prevailed through the 1970s, let us examine the controversy which arose around the role of finance capital in the monopoly capitalist city. Michael Harloe, a leading Marxist urban sociologist and long-time editor of the *International Journal of Urban and Regional Research*, defined one side of the debate:

> In general Harvey's emphasis on the role of finance rather than productive capital in cities has been criticized. He does not go quite so far as Lefebvre who sees finance capital, i.e. that concerned with circulation rather than production, becoming the dominant force in society, and urban conflicts, based on the role of such capital in property speculation and land, supplanting workplace conflict – a theory heavily criticized by Castells. However, Harvey distinctly sees this as a possibility, but his critics suggest that finance capital must remain secondary to productive capital because it ultimately has to abstract its wealth from surplus value and so is subordinate to the latter in the last analysis. (Harloe, 1976, 25)

This compulsion to assert the ultimate dominant, the great bane of categorical versus dialectical logic, became epitomized in all its structuralist simplicity in Harloe's conclusion that while finance capital 'can apparently be in control, the dominant role still falls to productive capital'. By implication, Lefebvre is thus an irreconcilable deviant who has forgotten the catechism of Marx, Harvey is unacceptably ambivalent and confused, while production reigns supreme and inviolable.

The major question, however, is not whether finance capital dominates industrial capital 'in the last analysis', but how it relates, as one fraction of capital, to other capital fractions within specific social formations, and how this affects class action. The issue is thus a conjunctural one, referring to the ensemble of class relations arising at particular places during particular periods of time. To reduce Marxist analysis to the assertion of ultimate structural determinations is to eliminate all historical and geographical specificity – and hence to eliminate the city itself as a subject of analysis.[1]

---

1. This is precisely what many urban analysts were led to conclude in the 1970s debate on the 'specificity of the urban'. The city was neither a 'theoretical object' nor a 'real object', with its own autonomous rules of formation and transformation. Marxists could study cities as expressions of broader and deeper societal structures but not as fields of significant structuration in themselves. To assign a theoretical and political specificity to the urban would be to fall back into the ideological trap of the Chicago School of urban ecology, as Castells claimed that Lefebvre had done. Few could see that what was being asserted, by Lefebvre and eventually by Harvey, was a more encompassing *spatial* specification of the urban. The urbanization process, far from being autonomous, was an integral part of the enveloping and instrumental *spatialization* that was so essential to the historical development of capitalism, a spatialization that was almost invisible to Marxism and other critical perspectives throughout most of the twentieth century.

How then should we assess finance capital? Although Lefebvre clearly stressed its role in the realization of surplus-value, he did not unambiguously assess its relations with other forms of capital, leaving open the old question of whether finance and industrial capital are in permanent conflict or, as Lenin argued in his theory of imperialism, they have virtually coalesced. In contrast, David Harvey appeared to have accepted both positions without regard to their seeming incompatibility. At times, he viewed finance capital as a separate fraction, a parasitic monopoly sector sucking up funds which would otherwise be available for housing and basic social services, and in direct conflict with industrial capital (from which it is able to extract part of the surplus in the form of rent and interest). At other times, Harvey treated finance capital almost in the Leninist sense as commingled with industrial capital in a monopoly imperialist union, co-operatively controlling the totality of the production process.

In an interesting paper on rent theory and working-class strategy, subjects which were directly linked to the debate on finance capital, Matthew Edel commented on Harvey's apparent duplicity:

> The two views of finance capital are not really compatible, and they lead to different strategic implications for the working class. If the first definition holds, and if financial monopolies or financial discrimination are making housing finance scarce, either for workers generally or for specific ghetto subgroups, then an anti-monopoly strategy – involving anti-redlining laws, state housing banks, or even moratoria on mortgage payments – would make sense. If, however, definition two applies and one integrated financial group controls both housing and employment, then a reduction of housing costs might be met by reduced wages. Finance capital's 'certain indifference as to whether accumulation takes place by keeping wages down in the immediate production process or by manipulation in the consumption sector', if it emerges as Harvey suggests it might, would create a need for the working class to organize simultaneously as wage earners and as consumers, in order to make any economic gains under capitalism. But such victories would be difficult, since tactical errors on the consumption struggle side would adversely affect the wage struggle, and vice versa. (Edel, 1977, 12)

Edel concludes, however, that the nature of the relationship between financial and industrial capital must be seen as historically determined and thus clearly subject to change. During the early phases of stagnation in the 1970s, financial barriers had a major effect. They were largely responsible for the transformation of metropolitan cores, through urban renewal and reduced low-income housing supplies, into spectacular new centres for finance capital institutions. But deepening economic crisis shifted the burden back to the reduction of real wages, thereby establishing the need for a two-front struggle over consumption and production relations simultaneously. Thus, Edel argued, an attack on only one sector of capital, be it banking or landlords, is no longer sufficient. This argument, however, needs to be elaborated further.

'Unproductive' finance capital has become an important element in the structure of contemporary capitalism, but not because it has supplanted industrial capital in the realization of surplus value. It obtains its importance from the increasing absorption of surplus value in collective consumption, in the reproduction of labour power and the social order, in the sustaining social production of urbanized space. Under competitive industrial capitalism, the organization of urban space for both production and consumption could be left largely to market forces, private property regulations, and normal competition among industrial producers for access to labour, materials, and infrastructure. The industrial capitalist city was primarily a production machine and as such took on a remarkably uniform spatial structure – the structure depicted so perceptively by Engels for Manchester and later by the urban ecologists for Chicago and other North American cities.

Finance capital under these conditions was relatively unimportant as a direct actor in the local urban context. Its primary role in capitalist spatialization was more global, shaping the imperialist expansion around the turn of the last century as the urban production machines surpassed their capacity to consume their product and triggered falling rates of profit and rising class conflict. Expanded reproduction on a global scale and the concurrent growth of monopoly capitalism, however, massively intensified the concentration of capital in the centres of advanced industrial countries and created increasing pressures for infrastructural investments, improved housing and services, and reinvigorated methods of social control. More than ever before, there was a need to intervene to reorganize urban space and to make urban systems function more effectively for the accumulation of capital and the management of social unrest. This brought finance capital more directly into the planning of urban space.

What began in the early part of the century was accelerated in response to the Great Depression and the Second World War, after it became clear that imperialist expansion and corporate monopoly alone would not eliminate class conflict and economic crisis. As a result of another round of restructuring, finance capital became still more significant in shaping urban space, in conjunction not only with industrial capital but increasingly with the other key agency of regulation and spatial restructuring, the state. This coalition of capital and the state worked effectively to replan the city as a consumption machine, transforming luxuries into necessities, as massive suburbanization created expanded markets for consumer durables. The growth of this 'crabgrass frontier' of tract homes also intensified residential segregation, social fragmentation, and the occupational segmentation of the working class.

It became increasingly clear to Marxist urban analysts in the 1970s, however, that this depression-induced urban restructuring, after playing a significant role in sustaining the post-war economic boom, was itself beginning to generate new forms of contradiction, crisis, and social struggle. Castells (1976) vividly captured the new urban spatial problematic that was developing in his description of *la ville sauvage*, the 'wild city', but the decade's most cogent theoretical interpretation, which would reverberate well into the 1980s, came from David Harvey as part of his overformalized but incisive analysis of 'The Urban Process Under Capitalism'. After tracing the flow of capital through three interweaving circuits of space-building investment, Harvey offers a most provocative conclusion:

> Capital represents itself in the form of a physical landscape created in its own image, created as use values to enhance the progressive accumulation of capital. The geographical landscape which results is the crowning glory of past capitalist development. But at the same time it expresses the power of dead labour over living labour and as such it imprisons and inhibits the accumulation process within a set of specific physical constraints.... Capitalist development has therefore to negotiate a knife-edge path between preserving the exchange values of past capital investments in the built environment and destroying the value of these investments in order to open up fresh room for accumulation. Under capitalism, there is then a perpetual struggle in which capital builds a physical landscape appropriate to its own condition at a particular moment in time, only to have to destroy it, usually in the course of crises, at a subsequent point in time. The temporal and geographical ebb and flow of investment in the built environment can be understood only in terms of such a process. (Harvey, 1978, 124)

Here the city, the urban built environment, is embedded in the restless geographical landscape of capital, and specified as part of a complex and contradiction-filled societal spatialization that simultaneously enhances and inhibits, provides new room and imprisons, offers solutions but soon beckons to be destroyed. The history of capitalism, of urbanization and industrialization, of crisis and restructuring, of accumulation and class struggle, becomes, necessarily and centrally, a localized historical *geography*. This capsule of insight marked the end of Harvey's ambivalence and the opening of a new phase in Marxist urban analysis that will be examined in Chapter 7.

## The Regional and International Spatial Problematic

Marxist urban analysis and Marxist interpretations of regional and international development rarely overlapped during the 1970s. Although a
new regional political economy seemed to be taking shape alongside the
already well established urban political economy, there was relatively
little sharing of ideas.[2] Instead of building upon the urban debates and
arguments, regional analysis drew primarily from neo-Marxist theories
of international development and underdevelopment, which had from
the 1960s begun to incorporate a challenging, if somewhat serendi-
pitously discovered, spatial perspective.

Inexorably, their work also attracted the familiar charges of spatial
fetishization and submergence of class analysis. One of the most
vehement attacks came from an economist, Ann Markusen, who was
then just beginning her work in regional political economy. Markusen
was especially critical of 'younger Marxist social scientists, who unwit-
tingly mimic the tendency within bourgeois social science to assign
characteristics to places and things, rather than sticking to the dynamics
of a process as the analytical focus (1978, 40). The process dynamics
were essentially historical and political, she claimed, not geographical.
Geographies, regional inequalities and regionalisms, came later as
outcomes of history and politics, as products of historically not geographically uneven development. The nineteenth-century ghosts of
historicism had spoken again through another of their pious descendants.

Almost no one, therefore, seemed to notice the argument that was
developing since the early 1960s in the work of Ernest Mandel, one of
the foremost Marxist political economists of the twentieth century. He
published in 1963 what can be referred to today as the first major
empirical study in regional political economy, prefiguratively titled 'The
dialectic of class and region in Belgium'.[3] Later, in his influential *Marxist
Economic Theory* (1968), he noted that uneven regional development
is 'usually underestimated in Marxist economic writing' yet is 'one of the
essential keys to the understanding of expanded reproduction' (Volume
1, 373). He would push his argument still further in his later works,
affirming, as we have seen in his most emphatic statement (Mandel,

---

2. The most notable exceptions were the Special Issue on Regional Uneven Development of *The Review of Radical Political Economics* (Volume 10, 1978); Doreen Massey's
survey of Marxist regional analysis in *Capital and Class* (Volume 6, 1978); and Alain Lipietz,
*Le Capital et son espace* (1977).

3. The article appeared in *New Left Review* 20, 1963, 5–31, one of the very few
explicitly regional or spatial analyses ever to be published by what has been for many years
the anglophonic Left's most widely circulated journal.

1976) that the unequal development of regions and nations is as fundamental to capitalism as the direct exploitation of labour by capital.

The writings of Immanuel Wallerstein on the capitalist 'world system' and its characteristic structure of centres, peripheries, and semi-peripheries; André Gunder Frank and the Latin-American 'Structuralist' school on metropolis-satellite relations, underdevelopment, and dependency; Samir Amin on the accumulation of capital on a global scale and the distinctive properties of central and peripheral capitalism; Arghiri Emmanuel on unequal exchange and the transfer of value – all implicitly complemented and expanded Mandel's arguments but rarely looked directly at the powerful spatializations of both Marxist theory and the concrete history of capitalist development suggested in his interpretation of *Late Capitalism*. This is why I describe their identification of a spatial problematic at the regional and international scale as 'serendipitous', an almost accidental adjunct to their analyses of expanded reproduction under capitalism. But let us dig deeper into the implicit problematic that resulted from this increasing spatialization of regional and international development theory in the 1970s.

### On the necessity of geographically uneven development

The spatial problematic and its socio-political ramifications on the regional and international scales hinge upon the importance assigned to geographically uneven development in the genesis and transformation of capitalism. Stated somewhat differently, it depends upon the degree to which Mandel's assertion about the interpretive equivalence of unequal development and the capital-labour relation can be accepted and built upon in Marxist theory and political practice. The question is clear: is geographically uneven development a necessary as well as contingent feature of capitalism?

To answer this question, it is useful to begin by distinguishing between an analysis of the general laws of motion of capital under pure and homogeneous (that is to say, spatially undifferentiated) conditions; and a more concrete analysis of the conditions which exist within and between particular social formations. In the former, abstracted from the particularities of time and place, the fundamental logic of structure and process can indeed be viewed as spaceless. Following Marx in *Capital*, the intrinsic nature of capitalism can be revealed as a system of production built upon a set of internal contradictions, beginning with that between the forces and relations of production, which must ultimately lead, through crisis and class struggle, to revolutionary transformation. In this essentially theoretical and abstracted interpretation of capitalism, with its tacit parallels to Western conceptualizations of science based on general laws freed from the specificity of time and place, geographically uneven development is not only irrelevant, it is defined away, logically obliterated.

Marxist analyses built upon these general laws of motion and centralized around the contradiction between labour and capital in the production process effectively demonstrate the ephemerality of capitalism, its imminent self-destruction. But they do not, by themselves, explain how and why capitalism has continued to exist. To do this requires more direct attention to reproduction processes, especially to the particularities of the expanded reproduction of capitalism. When the general laws of motion of capital are grounded in the concrete history and geography of the capitalist mode of production – a conjunction which Mandel claims Marxist theory has never satisfactorily clarified – the constitutive role of geographically uneven development becomes more visible.

There are many parallels between this argument and that presented earlier regarding the urban spatial problematic. In both, the vital spatiality of capitalism is characteristically defined away by appeal to orthodox Marxist theory in its most abstracted form. But when concretized in the context of expanded reproduction, in the specificity associated with the 'intensification' and 'extensification' of capitalism, the significance of spatiality is reasserted. Again, the leading question is not why capitalism cannot last, but why and how it has survived from the competitive industrial form of Marx's time to its contemporary metamorphosis – the question which permeates Lefebvre's work over the past fifty years.

How then can we define the role of geographically uneven development within this framework of expanded reproduction? We must begin again with Lefebvre's generative assertion that capitalism has been able to survive and achieve 'growth' by producing and occupying a space, by a pervasive and problem filled spatialization process. To this we add a Mandelian specification that the survival of capitalism and the associated production of its distinctive spatiality have depended upon the differentiation of occupied space into 'overdeveloped' and 'underdeveloped' regions, the 'juxtaposition and constant combination of development and underdevelopment'. How this specific geography of capitalism takes shape, and is reshaped over time, is just beginning to become clear.

Regional underdevelopment is an integral part of extended or expanded reproduction, creating large reservoirs of labour and complementary markets capable of responding to the spasmodic and contradictory flow of capitalistic productivity. 'If in order to survive', Mandel argues, 'all of the working population found jobs in the region where they lived, then there would no longer be reserves of wage labour free for the sudden expansions of industrial capitalism' (1976, 43). When these regional labour reserves[4] are not created by 'natural' population movements, they are produced through direct force and other means. When they dry up in one area, they are created again in another.

The creation of labour reserve space also supplies important markets for capital in the 'overdeveloped' centres of accumulation. This combination of economically depressed labour reserves serving also as markets for the surplus product of the centre is maintained and intensified through a system of unequal exchange, or what can be more broadly termed the geographical transfer of value. As Mandel suggests, this unequal exchange is itself based on regional differentiation.

> If the rate of profit were always the same in all regions of a nation and in all countries of the world, as well as in all the branches of industry, then there would be no more accumulation of capital other than that made necessary by demographic movement. And this itself would be modified in its own turn by the impact of the severe economic stagnation that would ensue. (Mandel, 1976, 43)

In *Late Capitalism* (1975), Mandel presents an innovative historical synthesis of the importance of geographically uneven development and the relation between the geographical and sectoral transfers of value.[5] He builds this synthesis upon a critical contradiction in the process of expanded reproduction, that between the differentiation and equalization of the rate of profit (and such related measures as the organic composition of capital and rates of exploitation). Whereas normal capitalist competition tends toward an equalization in the rate of profit between sectors and among regions, expanded reproduction must feed off the extraction of 'super-profits' (that is, higher than the average rate), which in turn requires sectoral and/or regional differentiation. Thus, Mandel explains, the actual growth process of capitalism never

---

4. Regional labour reserves may also be located within urban areas. Geographically uneven development, as Mandel notes for the entire capitalist system, 'appears as a hierarchical structure of different levels of productivity', from the local to the global. This multi-scalar hierarchy is itself produced by capitalist spatialization.

5. Mandel never specifically uses the terms 'geographically uneven development' or the 'geographical transfer of value'. As far as I can tell, they were first used consistently in my writings to describe, with a more explicit geographical emphasis, what Mandel usually called unequal development and unequal exchange between regions and nations. In the late 1970s, I worked closely with a Greek architect/planner, Costis Hadjimichalis, as he developed his own ideas on 'The Geographical Transfer of Value', the title of his unpublished dissertation (Graduate School of Architecture and Urban Planning, UCLA, 1980). For an extension of this work by Hadjimichalis, which had a strong influence on my thinking at the time, see his recently published book, *Uneven Development and Regionalism* (1986).

achieves the full equalization of rates of profit. It is always, sectorally and spatially, unevenly developed even, he notes, in the 'ideal case' of a homogeneous beginning.

> *The accumulation of capital itself produces development and underdevelopment as mutually determining moments of the uneven and combined movement of capital.* The lack of homogeneity in the capitalist economy is a necessary outcome of the unfolding laws of motion of capital itself. (Mandel, 1975, 85; emphasis Mandel's)

What I am arguing here through the words of Mandel is not simply that capitalist development is geographically uneven, for some geographical unevenness is the result of every social process. The key point is that capitalism – or if one prefers, the normal activity of profit-seeking capitalists – intrinsically builds upon regional or spatial inequalities as a necessary means for its continued survival. The very existence of capitalism presupposes the sustaining presence and vital instrumentality of geographically uneven development.

What is it that becomes geographically differentiated? The list begins with rates of profit, a crucial parameter, but also includes the organic composition of capital, labour productivity, wage rates, the costs of materials necessary for the reproduction of labour power, levels of technology and mechanization, the organization of labour, and the incidence of class struggle. These differentials are maintained through geographically and sectorally uneven allocations of capital investment and social infrastructure, the locational concentration of centres of control over labour and the means of production, the interlocking circuits of capital in the urbanization process, and the particular forms of articulation between capitalist and non-capitalist relations of production. All form part of the complex and distinctive spatialization that has marked the historical development of capitalism from its origins.

At the same time, there is also a persistent tendency towards increasing homogenization and the reduction of these geographical differences. This dialectical tension between differentiation and equalization is the underlying dynamic of geographically uneven development. It is a primary source of the spatial problematic at every geographical scale, from the immediacies of everyday life at the workplace, the household, and the urban built environment to the more distant structure of the international division of labour and the capitalist world economy. It is also embedded, although usually unseen, in Marx's treatment of primitive accumulation and the antagonism between town and countryside, as well as in more contemporary Marxist debates on the theory of the state.[6] Alain Lipietz captures the connection with the role of the state most vividly:

> Faced with the uneven development of socio-economic regions, the state must take care to avoid sparking off the political and social struggles which would arise from too abrupt a dissolution or integration of archaic modes of production. This is what it does in a general fashion when it inhibits the process of articulation (protectionism) or when it intervenes promptly to remove social consequences (permanent displacement allowances). But as soon as internal and international evolution make it necessary, capitalist development assigns to the state the role of controlling and encouraging the establishment of a new inter-regional division of labour. This 'projected space' comes into more or less violent conflict with 'inherited space'. State intervention must therefore take the form of organizing the substitution of projected space for present space. (Lipietz, 1980, 74; trans. from 1977)

Here again is Harvey's restless geographical landscape, his knife-edge path between preservation and destruction, etched into a regional as well as urban problematic and similarly connected to that most powerful of capitalist territorialities, the national state.

It is no surprise, however, to find that the dynamic contradiction between spatial differentiation and equalization came to the surface most prominently in the 1970s with regard to the internationalization of capital, the scale at which the ultimate geographical limits of capitalism are reached. Geographically uneven development was an important issue in the early theories of imperialism, and continued to be noticeable as it became clearer that the geographical expansion of capitalism was not a simple process of homogenization, with the pure relations of *Capital* replacing pre-capitalist production relations like molasses spreading on a plate. Instead, the non-capitalist world became complexly articulated with the capitalist, its relations of production simultaneously and selectively being disintegrated and preserved. The resultant international division of labour, capitalism's distinctive form of global spatialization, grows out of this contradiction-filled combination

---

6. Critical regional planning theorists saw a somewhat similar logic to the dynamics of national development. For example, Gunnar Myrdal's model of initial locational advantage and cumulative causation, with its contrary 'spread' and 'backwash' effects, illustrated how development triggered tendencies toward both increasing and decreasing regional inequalities (Myrdal, 1957). These Myrdalian and related models (e.g., Hirschmann, 1958) were later creatively extended in John Friedmann's 'general theory of polarized growth' (1972), which inserted a more explicit political struggle into the dynamics of unbalanced regional growth and the core-periphery structure. In all of these planning approaches, however, the spatiality of development was seen primarily as an outcome of unproblemitized diffusion processes that were there to be innocently manipulated by benevolent state intervention.

of differentiation/equalization, disintegration/preservation, fragment-ation/articulation.

Christian Palloix, who with Aglietta and Lipietz was instrumental in the development of the French 'Regulation School' of political economy, explicitly focused on these contradictions of geographically uneven development in the internationalization of capital – or what he called 'the spatial extension of the capitalist mode of production'.[7]

> The self-expansion of capital on a world scale supports the international spread of branches of production and the tendency to uniformity seems dominant. This fails to take into account the fact that differentiation occurs when a branch of industry internationalizes. It disappears and is eliminated in one place in such a way that *new conditions* of production and exchange are created, giving rise to differentiation.... The essence of the international-ization of capital is that *this tendency* toward equalization *is immediately checked by the differentiation of conditions of production and exchange.* Thus, in an important way, the tendency of uniformity leads to *the condition* for the differentiation. (Palloix, 1977, 3; emphasis in original)

> Thus, in its international aspect, capital prevents any unity of the international working class by dividing up different working classes, taking advantage of areas of uneven development and amplifying existing schisms. The inter-nationalization of capital is antagonistic to the international struggle of the proletariat which attempts to reestablish the unity of the working class. (Ibid., 23)

### The core-periphery debate

Anglophonic interpretations of the 'international aspect' of capitalist development were dominated in the 1970s by theories of underdevelop-ment and dependency and Wallerstein's work on world systems. There were important French connections in this literature (for example, Wallerstein's Braudelian inspirations), but the French explicitness with space and spatiality tended to fade away in English translation. Alter-natively, the somewhat disguised geographies that came through provoked harsh reactions from anglophonic Marxist historians, econ-omists, sociologists, and even some geographers.

Virtually everyone seemed to accept that there existed a core and periphery structure to the capitalist world economy; that there were

---

7. Again the French seem so much more comfortable with explicitly spatial termin-ology. In all the work I have read on the internationalization of capital written before 1980 in English, I have not found anyone willing to describe what they were studying as the 'spatial extension' of capital.

certain 'core' countries which were primary centres of industrial production and accumulation and a subordinate, dependent, and highly exploited set of 'peripheral' countries forming part of a 'Third World'. But in the absence of a rigorous conceptualization of geographically uneven development, the theoretical and political significance of this core-periphery structure was difficult to understand, especially when only the top (world system) level of the complex scalar hierarchy was seen as explaining the history of capitalist accumulation and class struggle. This seemed to explain capitalism 'from above' rather than 'from below', externally rather than internally, where the empirically-oriented Marxist historians were most comfortable.

To many Marxist scholars, therefore, the neo-Marxist school of underdevelopment and dependency theorists appeared to be de-emphasizing class and placing the causes of poverty, exploitation and inequality almost exclusively in the hands of the core countries and a fixed international division of labour based on some form of colonial or neo-colonial domination. Exploitation, indeed the entire capitalist system, was being viewed as arising from a geographical process between core and peripheral areas that operated mainly at the level of relations between nations. In defence of his views, André Gunder Frank declared that 'dependency theory is dead – long live dependency and class struggle,' demonstrating slyly that he had not forgotten what comes first. But there was little attempt to explain in more detail the vital connection that was being made in dependency theory and in the structural polarization of core versus periphery between the social and spatial dimensions of capitalism.

Wallerstein, however, went further. He described the capitalist world system as revolving around two basic 'dichotomies', one of class (bourgeois versus proletarian) and the other of 'economic specialization' within a spatial hierarchy (core versus periphery), adding later a 'semi-periphery' to roughly parallel the formation of a 'middle' class. He then added:

> The genius, if you will, of the capitalist system, is the interweaving of these two channels of exploitation which overlap but are not identical and create the cultural and political complexities (and obscurities) of the system. Among other things, it has made it possible to respond to the politico-economic pressures of cyclical economic crises by re-arranging spatial hierarchies without significantly impairing class hierarchies. (Wallerstein, 1976, 350–1)

This double dichotomy pushes beyond conventional underdevelopment theory and up to the edge of an assertive socio-spatial dialectic. But the push is weakened somewhat by Wallerstein's failure to theorize the spatial structure 'at the same level' as the social, a lost opportunity to make the core–periphery relation more than an evocative descriptive metaphor.[8]

To do so required a conceptualization of the spatial hierarchy of cores and peripheries as both the product and the instrumental medium of geographically uneven development. As such, the core-periphery structure becomes fundamentally homologous to the vertical structure of social class. Not only do they interweave and overlap, they originate from the same sources and are shaped by the same contradictory relations of capital and labour that define the capitalist mode of production itself. The formation of core and periphery is thus inseparable from the formation of an industrial bourgeoisie and an urban proletariat.

Continuing this Marxist spatialization, both the social and spatial structures are constituted around an exploitative relation rooted in control over the means of production and sustained by an appropriation of value by the dominant social class. Class struggle arises from increasing consciousness of the doubly exploitative nature of the social and spatial structure and must be aimed at the simultaneous transformation of both, for they each can be re-arranged to dampen the effects of struggle in the other. This dual grounding may be the real, but rarely seen, 'genius' of the capitalist system, the most important 'discovery' of the 1970s.

It must be remembered that this double polarization occurs at many different levels or scales, and that at each scale the specific social and spatial configurations can change over time, without necessarily changing the underlying structure itself. A core country or core region within a country can become part of the periphery and an erstwhile periphery can become part of the core, just as individuals and families can shift from proletariat to bourgeoisie from one generation to another. The polarized structure at all scales is also blurred and mystified in its material forms, giving to actual social and spatial divisions of labour a more complex and finely grained segmentation. The international division of labour, the state-bounded inter-regional division of labour, the urbanized division of labour within metropolitan areas, the division of labour in the smallest locality, factory, or household, these are not captured in all their complexity by the core-periphery structure. But it is *there* nonetheless, insofar as the social and spatial relations are linked into capitalist production. The specific geography of capitalism can be restructured but

---

8. For the most part, Wallerstein's spatial assertions were ignored by Marxist and other historians and sociologists, except to provide 'proof' that he had reified the capitalist world system into a new (geographical) determinism, the familiar over-reaction of a hyperactive historical imagination.

is never unstructured or entirely free from a fundamental socio-spatial polarization.

The mutable spatial hierarchy of core-periphery relations is still poorly understood and has created widespread misapprehension. That some peripheral countries are currently rapidly industrializing while once advanced industrial regions are declining, for example, does not mean that underdevelopment theory, the existence of a global core-periphery structure, and the necessity for geographically uneven development have somehow to be rejected. There is still the same 'juxtaposition and constant combination of development and underdevelopment', homogenization and differentiation, disintegration and preservation, amidst the changing empirical patterns.

Another misunderstanding caused a stir in the 1970s and slowed the advancement of a Marxist regional and international political economy. It revolved around the question of whether regions exploit other regions, especially given the assertion of an interactive relation between core and periphery on the one hand and bourgeoisie and proletariat on the other. Were there bourgeois and proletarian regions, and if so wasn't this the ultimate in reification? Surely only people exploit other people, class exploits/dominates class? To respond to these questions – and to summarize many of the arguments that have already been presented – it is necessary to turn to the last of the debates which will be reviewed here.

### On the geographical transfer of value

Once put into explicit focus, the notion of 'unequal exchange' (Emmanuel, 1972; Amin, 1973 and 1974; Mandel, 1975) or, to give it a more inclusive definition, the geographical transfer of value (Hadjimichalis, 1980), seems exquisitely obvious from the viewpoint of Marxist political economy, a straightforward derivation from the general discussion of the transfer of value found in *Capital*, Volume III. As Marx explained, the differentiation between firms, branches, and sectors of capitalist production in terms of such key measures as the organic composition of capital and labour productivity, combined with the omnipresence of competition, leads to a transfer of value measured in terms of socially necessary labour time. Through the operations of a competitive market, a relatively efficient (and usually more mechanized) firm, for example, will get more value from the exchange process than it actually produces through the labour process, based upon little more than the differences between embodied labour value and the prices of production across all competing firms. Similarly, the less efficient firm would receive less value.

All that is necessary to define a geographical transfer of value is to give capitalism a concrete geography, to move production and exchange off the spaceless head of a pin and into a differentiated and unevenly developed spatiality. In the geographical landscape of capital, there will be areas of greater and lesser productivity, high and low capital to labour ratios (that is to say, organic composition), varying rates of profit all within the same integrated market for commodities. Market exchange thus becomes a vehicle not only for the transfer of value between firms and sectors but also for a geographical transfer of value. Certain areas will experience a net gain in value terms while others incur a net loss, and this will have some effect on the geography of accumulation, on the formation of centres and peripheries at every spatial scale.

It is also conceivable that the specific elements which determine the value content of aggregate production in different geographical areas will be so distributed and mixed across firms and sectors that they cancel themselves out, so to speak, leading to no significant areal variations. A broad spatial equalization in profit rates, organic composition, productivity and related indices would prevail and the geographical transfer of value would be of little consequence. But this conceivable spatiality of capitalism is an idealized equilibrium that rarely if ever occurs in the persistently disequilibrating real world of capitalist development. It may exist for a period among a specific set of regions, it may even have existed at times among, say, the regions of the United States. But tendencies towards equalization are always accompanied by tendencies towards differentiation and hence towards the production and reproduction of geographically uneven development. And when there is geographical unevenness in capitalist development, there will be geographical transfers of value. The net flow between specific regions may not remain constant over time and can even reverse direction, but the inducing juxtapositions of development and underdevelopment still remain somewhere in the system. The mutable spatiality of the geographical transfer of value is thus very much like that of the core-periphery structure and the multi-layered empirical mosaics of geographically uneven development with which it is tightly intertwined.

More specifically defined, the geographical transfer of value is the mechanism or process through which a part of the value produced at one location, area, or region is realized in another, adding to the receiving region's localized accumulation base. This transfer operates at two levels. The first, and 'deepest' in terms of the spatial structure, is the transfer of value arising from geographical differences in production systems and associated labour processes – the simple spatialization of Marx's transfer of value between firms and sectors. This movement is invisible empirically and thus not conveniently susceptible to direct quantitative measurement. It derives primarily from the political logic of the labour theory of value and thus exists intrinsically and qualitatively within capitalist exchange relations (at least insofar as the labour theory is adhered to).

Following Hadjimichalis (1980), this basic transfer can be crudely approximated in the statistical difference between the total value added in production for a given region and the total income received for this production by the region's population, assuming other complicating flows and transfers can be accounted for. At the very least, this gives a rough indication of the likelihood that the value transfer is markedly positive or negative, although even here the estimation is complicated by the many channels of value transfer which operate at the more readily apparent second level.[9]

The transfer of value at this second level takes many different forms, is more measurable, and can be more easily modified and regulated. Michael Kidron (1974), in a critique of Emmanuel's theory of unequal exchange, described what I am calling the second level of the geographical transfer of value in conjunction with an analysis of the centralization and concentration of capital. 'Positive centralization' in Kidron's terms is essentially equivalent to primitive accumulation and involves a net gain to a 'centre' without a direct value loss to the not fully capitalist social formations in the 'periphery', such as occurs through colonial tax tributes and plunder.[10] 'Negative centralization', in contrast, magnifies centre–periphery differences in the relative sizes of capital through the destruction of surplus in the periphery through war and 'waste' – in other words, the diversion of the peripheral surplus into such non-productive activities as the production and/or purchase of armaments. Finally, 'neutral centralization' involves the transfer of surplus between capitals with no change in the total produced value. This covers various processes: the 'brain drain' of selective migration; the net export of profits, fees, royalties, and so on, over 'aid' receipts; the transfer of direct control over peripheral capital to multinational corporations

---

9. Hadjimichalis (1980 and 1986) used this measure in a set of regional maps of Spain, Italy, and Greece over several time periods. The core regions of Madrid, the industrial triangle of northern Italy, and especially the Athens area remained remarkably stable over time as net 'receivers' in the value exchange, but many other regions shifted from receivers to 'donors', or vice versa.

10. Mandel argues that this was the chief form of metropolitan exploitation of the Third World prior to the Second World War. See his interesting chapter in *Late Capitalism* (1975) on 'Neo-colonialism and Unequal Exchange', which contains another but more sympathetic critique of Emmanuel (1972). It should also be noted that this 'positive centralization' runs through Marx's depiction of the antagonism between town and countryside, the parent 'model' for the geographical transfer of value, the core-periphery relation, and the necessity for geographically uneven development in the history of capitalism.

based on core countries; the techniques of transfer pricing within multinational conglomerates (or multilocation firms); and, in particular, the amplified unequal exchange arising from marked differentials in the terms of trade in commodities, what Emmanuel, Mandel, Amin and others argued had become the most important basis for the geographical transfer of value in the post-war international economy.

In all these forms of geographical value transfer, the basic pattern is the same whether it is described as part of the centralization and concentration of capital or the domination of core over periphery. A portion of the surplus product generated at one location is blocked from being locally realized and accumulated, while the surplus produced at another location is augmented. There is some comparability here with the exploitation of labour by capital in the workplace. Both the social and spatial channels of exploitation are intrinsically rooted and sustained in the mode of production – so deeply that to eliminate them entirely requires a revolutionary and global transformation. However, both are also somewhat malleable, capable of being intensified or decreased by institutional or other means at specific times and places, without necessarily destroying the structural polarization. The two channels may appear separate, but they are intertwined; and both are filled with politics and ideology.

It should be remembered that Marx never argued that production, realization, and accumulation always occurred at the same place and thus kept open a revealing window on to the geographical landscape of capitalism and its spatial channels of exploitation. But a long tradition of obfuscating the material significance of capitalist spatiality prevented Marxists from seeing very clearly through this window until the 1970s. The first systematic sighting of the geographical transfer of value, to illustrate the point, may have been Arghiri Emmanuel's *Unequal Exchange: A Study of the Imperialism of Trade* (1972).

Emmanuel, however, saw unequal exchange operating almost entirely on the international level as a mechanism whereby the value produced in peripheral capitalist countries was transferred persistently to the core via the imperialism of trade and differential wages. For most of the decade, the debate on the geographical transfer of value remained concentrated around this focus. Very little was done on conceptualizing and empirically examining the more general and multi-scale processes behind the geographically uneven development of capitalism. The geographical imagination was awakening after a long sleep, but its vision remained blinkered and blurred.

There were many controversial issues which arose in the specific debates over unequal exchange at the global scale, but most tended to bury rather than elucidate the spatial implications of what was being argued. This was most vividly exemplified in the heated discussions among Marxists over the political questions raised by the core–periphery value transfer and the apparent depiction of regions exploiting regions. Amin captures the gist of these questions:

> If the relations between the center of the system and its periphery are relations of domination, unequal relations, expressed in a transfer of value from the periphery to the center, should not the world system be analyzed in terms of bourgeois nations and proletarian nations, to employ the expressions which have become current? If this transfer of value from the periphery to the center makes possible a large improvement in the reward of labor at the center than could have been obtained without it, ought not the proletariat at the center to ally itself with its own bourgeoisie to maintain the world status quo? If this transfer reduces, in the periphery, not merely the reward of labor but also the profit margin of local capital, is this not a reason for national solidarity between the bourgeoisie and the proletariat in their struggle for national economic liberation? (Amin, 1974, 22)

In many ways, this restated, with a few new twists, the same political and strategic issues that had split Marxist-Leninist policies towards national and international development since the beginning of the imperialist era. Amin's resolution of the dilemmas resurrected by the few new twists was revealing. No, he argued, we should not analyse the world system in terms of bourgeois and proletarian nations, although we must recognize the profound differences between central and peripheral capitalist social formations and the significance of the persistent transfers of value which flow between them. At the same time as we recognize this powerful spatial structure to international capitalism, however, we must reassert the dominance of the 'social plane'. For Amin, the key social contradictions of capitalism are 'not between the bourgeoisie and proletariat of each country considered in isolation, but between the world bourgeoisie and the world proletariat' (Amin, 1974, 24).

'Workers of the world system unite', is not a particularly new strategy, but its reaffirmation in this context served to deflect attention away from the more substantive rethinking of the 'social plane' that was necessary to accommodate, in theory and in practice, the only recently visible spatial dynamics of international capitalist development. Like so many otherwise perspicacious observers of the geographical landscape of capital in the 1970s, Amin reasserted the dominance of the social (and the historical) rather than confronting the revealed challenge of a socio-spatial dialectic and the double-edged struggle this implied. Mandel added greater clarity to the debate on unequal exchange and neocolonialism (1975, 343–76), but also left many key questions unasked and/or unanswered.

The question of whether regions exploit regions must therefore be answered affirmatively, but this affirmation can make theoretical and political sense only when regionalization and regionalism are seen as social processes hierarchically structured by the fundamental relations of production. Regions in this sense are people, classes, social formations, spatial collectivities, active and reactive parts of the geographical land-scape of capitalism. They may never be uniformly bourgeois or proletar-ian, but their localized production systems can function as either poles of accumulation or poles of devalorization with respect to other regions at the same scale, creating an interactive polarity that is homologous (but not equivalent) to the relation between the bourgeoisie and the proletar-iat.

If exploitation is of primary concern, then individual regions (as socio-spatial systems) must be seen as immersed in at least three chan-nels of exploitation. One is defined by the local relations between capital and labour in the production process, a second by the inter-regional relations within a larger spatial division of labour at a given scale, and a third by a nesting in a multi-scalar hierarchy of exploitative relations that extends from the global to the local, from the world system to the individual factory and household units. This comprises a much more complicated field of struggle against exploitation than conventionally envisioned, but no one ever said the struggle would be easy.

The first of these channels has traditionally so dominated the atten-tion of Marxists that the other two have remained much less visible over the past hundred years. And when the two spatial channels came more clearly into view in the 1970s, the orthodox were threatened while most of the insightful discoveries, although eliciting widespread enthusiasm, could not be effectively communicated without confusion, ambivalence, and exaggeration. A theoretically and politically challenging spatial problematic was bubbling to the surface, but the old historical lid was still on too tight.

# 5

# Reassertions:
# Towards a Spatialized
# Ontology

For its part, Marxist research has up to now ... considered that transformations of space and time essentially concern ways of thinking: it assigns a
marginal role to such changes on the grounds that they belong to the ideological-cultural domain — to the manner in which societies or classes *represent*
space and time. In reality, however, transformations of the spatio-temporal
matrices refer to the materiality of the social division of labour, of the structure of the State, and of the practices and techniques of capitalist economic,
political and ideological power; they are the *real substratum* of mythical, religious, philosophical or 'experiential' representations of space-time. (Poulantzas, 1978, 26)

This primal material framework is the mould of social atomization and splin-
tering, and it is embodied in the practices of the labour process itself. At one
and the same time presupposition of the relations of production and embodiment of the labour process, this framework consists in the organization of a
continuous, homogeneous, cracked and fragmented space-time such as lies at
the basis of Taylorism: a cross-ruled, segmented and cellular space in which
each fragment (individual) has its place, and in which each emplacement,
while corresponding to a fragment (individual) must present itself as homogeneous and uniform; and a linear, serial, repetitive and cumulative time, in
which the various moments are integrated with one another, and which is itself
oriented toward a finished product — namely the space-time materialized *par
excellence* in the production line. (Ibid., 64–65)

With the publication of *State, Power, Socialism* (1978), the last of his
major works, Nicos Poulantzas provided stirring evidence that he too
had discovered a socio-spatial dialectic of sorts and was seeking to redirect Marxist analysis toward a materialist interpretation of space and
time, an explicitly historical geography of capitalism. Building in part
upon the contributions of Lefebvre, he defined the spatial and temporal

*118*

'matrices' of capitalism, its material groundedness, as simultaneously presuppositions and embodiments of the relations of production. These intertwining matrices were not just the outcomes of a mechanical causality in which pre-existing relations of production give rise at some subsequent stage to a concrete history and a geography. 'Territory' and 'tradition', as Poulantzas labelled spatial and temporal matrices, were a *logical priority* (what Marx called *Voraussetzung*) and appear 'at the same time' as their presupposition. Neither were they to be seen, following the Kantian narrowing, only as ways of thinking, modes of representation. The creation and transformation of spatial and temporal matrices establish a primal material framework, the real substratum of social life.

These are complex and discomforting arguments which, like Lefebvre's introductory chapter in *La Production de l'espace*, assign to space what had so assertively been attached to time in the Marxist tradition: a fundamental materiality, a problematic social genealogy, a political praxis impelled through an indissoluble link to the production and reproduction of social life, and behind all this an ontological priority, an essential connection between spatiality and being. Without abandoning Marx, Poulantzas, again like Lefebvre, criticized Marxism for its persistent failure to see the material and ideological spatialization associated with the development and survival of capitalism, a spatialization intimately bound up with the social division of labour, the institutional materiality of the state, and the expressions of economic, political, and ideological power. At the same time he was clearly more comfortable with history than with the less familiar spatial matrix, a residual historical bias shared with Anthony Giddens and other social theorists who catch sight of the interpretive power of spatiality rather late in their careers. Nevertheless, Poulantzas joined with these other spatial envisionaries at the end of the 1970s in raising the call for a significant retheorization of the spatiality of social life.

Accompanying this call was another, more meta-theoretical project, a search for an appropriate ontological and epistemological location for spatiality, an active 'place' for space in a Western philosophical tradition that had rigidly separated time from space and intrinsically prioritized temporality to the point of expunging the ontological and epistemological significance of spatiality. Michel Foucault, an important contributor to this debate, recognized the philosophical invisibility of space relative to time (after being pushed to this realization by the interviewers from *Herodote*, op. cit.). His words are worth repeating: 'Space was treated as the dead, the fixed, the undialectical, the immobile. Time, on the contrary, was richness, fecundity, life, dialectic.' To recover from this historicist devaluation, to make space visible again as a fundamental referent of social being, requires a major rethinking not only of the concreteness of capitalist spatial practices but also of the philosophizing abstractions of modern ontology and epistemology.

## Materiality and Illusion in the Conceptualization of Space

The generative source for a materialist interpretation of spatiality is the recognition that spatiality is socially produced and, like society itself, exists in both substantial forms (concrete spatialities) and as a set of relations between individuals and groups, an 'embodiment' and medium of social life itself.

As socially produced space, spatiality can be distinguished from the physical space of material nature and the mental space of cognition and representation, each of which is used and incorporated into the social construction of spatiality but cannot be conceptualized as its equivalent. Within certain limits (which are frequently overlooked) physical and psychological processes and forms can be theorized independently with regard to their spatial dimensions and attributes. The classical debates in the history of science over the absolute versus relative qualities of physical space exemplify the former, while attempts to explore the personal meaning and symbolic content of 'mental maps' and landscape imagery illustrate the latter. This possibility of independent conceptualization and inquiry, however, does not produce an unquestionable autonomy or rigid separation between these three spaces (physical, mental, social), for they interrelate and overlap. Defining these interconnections remains one of the most formidable challenges to contemporary social theory, especially since the historical debate has been monopolized by the physical–mental dualism almost to the exclusion of social space.

The assertion of (social) spatiality shatters the traditional dualism and forces a major reinterpretation of the materiality of space, time, and being, the constructive nexus of social theory. In the first place, not only are the spaces of nature and cognition incorporated into the social production of spatiality they are significantly transformed in the process. This social incorporation–transformation sets important limits to the independent theorizations of physical and mental space, especially with regard to their potential applicability to concrete social analysis and interpretation. In their appropriate interpretive contexts, both the material space of physical nature and the ideational space of human nature have to be seen as being socially produced and reproduced. Each needs to be theorized and understood, therefore, as ontologically and epistemologically part of the spatiality of social life.

Conversely, spatiality cannot be completely separated from physical and psychological spaces. Physical and biological processes affect society

no matter how much they are socially mediated, and social life is never *except*
entirely free of such restrictive impingements as the physical friction of *digitally*
distance. The impress of this 'first nature' is not naively and independently given, however, for its social impact always passes through a
'second nature' that arises from the organized and cumulative application of human labour and knowledge. There can thus be no autonomous naturalism or social physics, with its own separate causal logic, in
the materialist interpretation of human geography and history. In the
context of society, nature, like spatiality, is socially produced and reproduced despite its appearance of objectivity and separation. The space of
nature is thus filled with politics and ideology, with relations of production, with the possibility of being significantly transformed.

Neil Smith captures the meaning of this social production of nature in
his recent work, *Uneven Development* (1984), revealingly subtitled
'Nature, Capital and the Production of Space'.

> The idea of the production of nature is indeed paradoxical, to the point of
> sounding absurd, if judged by the superficial appearance of nature in capital-
> ist society. Nature is generally seen as precisely that which cannot be
> produced; it is the antithesis of human productive activity. In its most immedi-
> ate appearance, the natural landscape presents itself to us as the material
> substratum of daily life, the realm of use-values rather than exchange-values.
> As such it is highly differentiated along any number of axes. But with the
> progress of capital accumulation and the expansion of economic development,
> this material substratum is more and more the product of social production,
> and the dominant axes of differentiation are increasingly societal in origin. In
> short, when this immediate appearance of nature is placed in historical
> context, the development of the material landscape presents itself as a process
> of the production of nature. The differentiated results of this production of
> nature are the material symptoms of uneven development. At the most
> abstract level, therefore, it is in the production of nature that use-value and
> exchange-value, and space and society, are fused together. (Smith, 1984, 32)

A similar argument can be made with respect to cognitive or mental
space. The presentation of concrete spatiality is always wrapped in the
complex and diverse re-presentations of human perception and cognition, without any necessity of direct and determined correspondence
between the two. These representations, as semiotic imagery and cognitive mappings, as ideas and ideologies, play a powerful role in shaping
the spatiality of social life. There can be no challenge to the existence of
this humanized, mental space, a spatialized *mentalité*. But here too the
social production of spatiality appropriates and recasts the representations and significations of mental space as part of social life, as part of
second nature. To seek to interpret spatiality from the purview of

socially independent processes of semiotic representation is consequently also inappropriate and misleading, for it tends to bury social origins and potential social transformation under a distorting screen of idealism and psychologism, a universalized and edenic human nature prancing about in a spaceless and timeless world.

Tied to this interpretation of the connections between physical, mental, and social space is a key assumption about the dynamics of spatiality and hence about the relations between (social) space and time, geography and history. Spatiality exists ontologically as a product of a transformation process, but always remains open to further transformation in the contexts of material life. It is never primordially given or permanently fixed. This may seem obvious when so simply stated, but it is precisely this transformative dynamic, its associated social tensions and contradictions, and its rootedness in active spatial praxis, that has been blocked from critical theoretical consciousness over the past hundred years. The spaces that have been seen are illusive ones, blurring our capacity to envision the social dynamics of spatialization.

## Misplacing spatiality: the illusion of opaqueness

A confusing myopia has persistently distorted spatial theorization by creating illusions of opaqueness, short-sighted interpretations of spatiality which focus on immediate surface appearances without being able to see beyond them. Spatiality is accordingly interpreted and theorized only as a collection of things, as substantive appearances which may ultimately be linked to social causation but are knowable only as things-in-themselves. This essentially empiricist (but also occasionally phenomenological) interpretation of space reflects the substantive-attributive structure (Zeleny, 1980) that has dominated scientific thought since the philosophy of the Enlightenment, a powerful heritage of objective naturalism to which spatial and social theorists have repeatedly appealed for both insight and legitimacy.[1]

From this myopic perspective, spatiality is comprehended only as objectively measurable appearances grasped through some combination of sensory-based perception (a purview developed by Hume and Locke and later revised and codified by Comte and others in a much less sceptical form of positivism); Cartesian mathematical-geometric abstractions (extended to manifold non-Euclidian variations); and the mechanical

---

1. Relevant here are the epistemological critiques of positivism arising from a group of realist philosophers and significantly influencing the contemporary retheorization of spatiality. See, for example, Bhaskar (1975 and 1979), Keat and Urry (1982), Sayer (1984), and, more directly applied to human geography, Gregory (1978).

materialism of a post-Newtonian social physics or a post-Darwinian sociobiology. A more contemporary cynosure of this spatial myopia, especially influential in the French philosophical and scientific tradition, was the *fin de siècle* figure of Henri Bergson. As we look back to the extraordinary devaluation and subordination of space (relative to time) that marked the last half of the nineteenth century, Bergson increasingly stands out as one of its most forceful instigators. For him, time, the vital realm of *durée* was the carrier of creativity, spirit, meaning, feeling, the 'true reality' of our world and our consciousness. Space, in the form of the categorizing intellect, was seen as orienting the mind to quantity (versus quality), measurement (versus meaning). Space is thus seen as pulverizing the fluid flow of duration into meaningless pieces and collapsing time into its own physical dimensionalities. As Lefebvre frequently notes – most recently in an interview published in *Society and Space* (Burgel et al., 1987) – this Bergsonian view 'throws all sins onto space' and rigidly separates space and time as science versus philosophy, form versus life, a vindictive dichotomization that would influence Lukács and so many other historicizers throughout the twentieth century.[2]

In all these approaches, spatiality is reduced to physical objects and forms, and naturalized back to a first nature so as to become susceptible to prevailing scientific explanation in the form of orderly, reproduceable description and the discovery of empirical regularities (largely in the spatial co-variation of phenomenal appearances). Such a short-sighted approach to space has proved productive in the accumulation of accurate geographical information and seductive as a legitimization for a presumed science of geography. It becomes illusive, however, when geographical description is substituted for explanation of the social production of space and the spatial organization of society, in other words, when geographical appearances are asserted as the source of an epistemology of spatiality. Yet, as I have noted before, this is precisely how space has been theorized in mainstream social science and in the theoretical geography which took shape in the 1960s. Even when a narrow empiricism or positivism is eschewed, the 'spatial organization of society' is made to appear socially inert, a deadened product of the ordering

---

2. Bergson's key works on space and time are his *Essai sur les données immédiates de la conscience* (1889; see also Bergson, 1910) and *Matière et mémoire* (1896). For a contemporary paean to Bergson and his followers and to the privileging of time and history against the evils of spatialization, see David Gross, 'Space, Time, and Modern Culture', published in *Telos* (Winter 1981–82), 'a quarterly journal of radical thought'. So intense is the urge to maintain the prioritization of time over space in this journal that the front cover and each page-top of the Gross article states the title as 'Time, Space and Modern Culture'!

discipline of the friction of distance, the relativity of location, the statistics of ecological covariation, and the axioms of geometry. Within this optical illusion, theories are constructed which always seem to mask social conflict and social agency, reducing them to little more than the aggregate expression of individual preferences which are typically assumed to be (naturally? organically?) given. Lost from view are the deeper social origins of spatiality, its problematic production and reproduction, its contextualization of politics, power, and ideology.

It is not surprising that the development and persistence of this illusion of opaqueness, with its submergence of social conflict in depoliticized geometries, has been interpreted (by those who can see beyond the myopic illusion) as an integral part of the evolution of capitalism. As one critic observed, 'Time and space assume ... that character of absolute timelessness and universality which must mark the exchange abstraction as a whole and each of its features' (Sohn-Rethel 1978, quoted in Smith, 1984, 74). Time and space, like the commodity form, the competitive market, and the structure of social classes, are represented as a natural relation between things, explainable objectively in terms of the substantive physical properties and attributes of these things in themselves, rather than as a 'continuous, homogeneous, cracked and fragmented space-time' rooted in the capitalist labour process, to recall Poulantzas' incisive words. Illusions of opaqueness have pervaded 'normal' science, both physical and social, throughout the past several centuries, obscuring from view the problematic historical geography of capitalism.

## Misplacing spatiality: the illusion of transparency

A similar interpretation can be made of a second source of illusion in the theorization of spatiality, one which evolves in a complex interaction with the first, often as its attempted philosophical negation.[3] Whereas the empiricist myopia cannot see the social production of space behind the opacity of objective appearances, a hypermetropic illusion of transparency sees right through the concrete spatiality of social life by projecting its production into an intuitive realm of purposeful idealism and immaterialized reflexive thought. Seeing is blurred not because the focal point is too far in front of what should be seen, inducing nearsightedness, but because the focal point lies too far away from what should be seen, the source of a distorting and over-distancing vision,

---

3. The continuing historical 'see-saw' between the dominant naturalist/scientific current of positivism and empiricism on the one side, and its 'anti-naturalist hermeneutic foil' on the other, is vividly described in Bhaskar (1979).

hypermetropic rather than myopic. The production of spatiality is repre-sented – literally re-presented – as cognition and mental design, as an illusive ideational subjectivity substituted for an equally illusive sensory objectivism. Spatiality is reduced to a mental construct alone, a way of thinking, an ideational process in which the 'image' of reality takes epistemological precedence over the tangible substance and appearance of the real world. Social space folds into mental space, into diaphanous concepts of spatiality which all too often take us away from materialized social realities.

The philosophical origins of this approach to theorization are probably Platonic. It was certainly boosted by Leibniz's assertion of the relativism of physical space, its existence as an idea rather than a thing. But its most powerful source of philosophical legitimacy and elaboration is Kant, whose system of categorical antinomies assigned an explicit and sustaining ontological place to geography and spatial analysis, a place which has been carefully preserved in a continuing neo-Kantian inter-pretation of spatiality. The Kantian legacy of transcendental spatial idealism pervades every wing of the modern hermeneutic tradition, infil-trates Marxism's historical approach to spatiality, and has been central to the modern discipline of geography since its origins in the late nine-teenth century. The vision of human geography that it induces is one in which the organization of space is projected from a mental ordering of phenomena, either intuitively given, or relativized into many different 'ways of thinking'. These ideas about space are then typically allocated to categorical structures of cognition such as human nature or culture at its most general, or biographical experience at its most specific, or alter-natively to 'science', to the Hegelian 'spirit', to the structuralist Marxist 'ideological-cultural domain', to an almost infinite variety of possible ideational compartments and sources of consciousness in-between.

Exemplifying just such a neo-Kantian cognitive mapping is Robert Sack's work, appealingly titled *Conceptions of Space in Social Thought* (1980). Although it propounds a realist philosophy, and ventures far beyond the limits of pure spatial subjectivity, Sack's analysis of the 'elemental structures' of 'modes of thought' is virtually divorced from the specific influence of 'socio-material conditions', a subject which Sack, in consummate neo-Kantian fashion, assigns for possible future research. Instead, he packages the conceptualization of spatiality into a neat categorical dualism that opposes the 'sophisticated-fragmented' spatial meaning associated with the arts, sciences, and contemporary industrial society; to the 'unsophisticated-fused' conceptualization of the primitive and the child, of myth and magic. In many ways, this dicho-tomization of spatial concepts parallels and can be mapped into such equally procrustean dualisms of social theory as *Gemeinschaft* and

*Gesellschaft*, mechanical and organic solidarity, Tradition and Modernity, the 'raw' and the 'cooked'.

Again, there are useful insights to be drawn from such approaches, but they also serve to reinforce a fundamental illusion that disguises the social production of space, especially when the ideation of spatiality is represented as a direct route towards understanding its problematic and politically-charged production and reproduction. Generic human conceptions of space may have some intrinsic and identifiable qualities worthy of further study, but if spatial fusion-fragmentation is indeed one of the elemental structures of social thought, it must be ground at all times in the material conditions of human life and not made to float as a timeless and placeless universal of human nature. Neither fusion nor fragmentation arise in this thin air of transcendental idealism.

As Lefebvre points out, and Poulantzas, Foucault, Giddens, Gregory, and others repeat in different ways, spatial fragmentation as well as the appearance of spatial coherence and homogeneity are social products and often an integral part of the instrumentality of political power. They do not arise from material spatiality or the mode of production in some simple, deterministic fashion, nor do they reflect back on society, once established, with simplistic determinacy of another kind. But conceptions or representations of space in social thought cannot be understood as projections of modes of thinking hypothetically (or otherwise) independent of socio-material conditions no matter where or when they are found, whether they emanate from the collective minds of a band of hunters and gatherers or from the institutionalized citizenry of the advanced capitalist state.

### Refocusing on the elusive spatiality of capitalism

As I have argued, a peculiar anti-spatialism, rooted in part in Marx's reaction to Hegel, has stubbornly screened Western Marxism from an appropriate materialist interpretation of spatiality. Marx treated space primarily as a physical context, the sum of the places of production, the territory of different markets, the source of a crude friction of distance to be 'annihilated' by time and the increasingly unfettered operations of capital. It can be argued that Marx recognized the opacity of spatiality, that it can hide under its objective appearances the fundamental social relations of production, especially in his discussions of the relations between town and countryside. He also approached, if not so directly, the basic problematic of the socio-spatial dialectic: that social relations are simultaneously and conflictually space-forming and space-contingent. The spatial contingency of social action, however, was reduced primarily to fetishization and false consciousness and never received from Marx an effective materialist interpretation.

Arguments about the spatial contingency of social relations, that these social relations of production and class can be reconfigured and possibly transformed through the evolving spatiality which makes them concrete, are still perhaps the most difficult part of the materialist interpretation of space for contemporary Marxist scholars to accept. A continuing aversion to any hint of a geographical determination of social life makes it difficult to see that spatiality is itself a social product that is not independently imposed and that it is never either inert or immutable. The geography and history of capitalism intersect in a complex social process which creates a constantly evolving historical sequence of spatialities, a spatio-temporal structuration of social life which gives form to and situates not only the grand movements of societal development but also the recursive practices of day-to-day activity – even in the least capitalist of contemporary societies.

The production of spatiality in conjunction with the making of history can thus be described as both the medium and the outcome, the presupposition and embodiment, of social action and relationship, of society itself. Social and spatial structures are dialectically intertwined in social life, not just mapped one onto the other as categorical projections. And from this vital connection comes the theoretical keystone for the materialist interpretation of spatiality, the realization that social life is materially constituted in its historical geography, that spatial structures and relations are the concrete manifestations of social structures and relations evolving over time, whatever the mode of production. To claim that history is the materialization of social life would cause little controversy, especially amongst Marxist scholars. It is virtually axiomatic to historical materialist analysis. But it is at this fundamental, axiomatic, and ontological level that spatiality must be incorporated as a second materialization/contextualization of social being. The constitution of society is spatial and temporal, social existence is made concrete in geography and history.

The first explicit and systematically developed assertion of this fundamental premise comes from Lefebvre in *La Production de l'espace* (1974):

> There remains one question that has not yet been posed: what exactly is the mode of existence of social relations? Substantiality? Naturality? Formal Abstraction? The study of space now allows us to answer: the social relations of production have a social existence only insofar as they exist spatially; they project themselves into a space, they inscribe themselves in a space while producing it. Otherwise, they remain in 'pure' abstraction, that is, in repre-

sentations and consequently in ideology, or, stated differently, in verbalism, verbiage, words. (152–53, author's translation)

This 'discovery' is the wellspring for the major reconceptualization of postmodern critical social theory that has advanced significantly in the 1980s. Lefebvre, Poulantzas, Giddens, and many others who have begun to reconceptualize the spatiality of social life share several additional specific emphases: the instrumental power which adheres to the organization of space at many different scales, the increasing reach of this instrumental and disciplinary power into everyday life as well as into more global processes of capitalist development, the changing and often contradictory roles of the state in these power relations wherever they are implanted.

Whether seen as the formation of a spatial matrix or as spatial structuration, the power-filled social production of space under capitalism has not been a smooth and automatic process in which social structure is stamped out, without resistance or constraint, onto the landscape. From its origins, the development of industrial capitalism was rooted in a conflict-filled attempt to construct a socially transformative and encompassing spatiality of its own. The arenas of struggle were many: the destruction of feudal property relations and the turbulent creation of a proletariat 'freed' from its former means of subsistence; the related uprootings associated with the spreading enclosure and commodification of rural and urban land; the expansive geographical concentration of labour power and industrial production in urban centres (and the attendant if incomplete destruction of earlier forms of urbanization, industrialization, and rural life); the induced separation of workplace and residence and the equally induced patterning of urban land-uses and the built environment of urbanism; the creation of differentiated regional markets and the extension of the territorial role of the capitalist state; the beginnings of an expansion of capitalism on to a global scale. As Poulantzas argued (and Michel Foucault took up with such detail and insight in his *recherches*), 'the direct producers are freed from the soil only to become trapped in a grid – one that includes not only the modern factory, but also the modern family, school, army, the prison system, the city and national territory' (Poulantzas, 1978, 105). Poulantzas goes on to describe the paradoxical fragmentation and homogenization, spatial differentiation and levelling-off of differences, that lies behind the geographically uneven development of capitalism:

Separation and division in order to unify; parcelling out in order to encompass; segmentation in order to totalize; closure in order to homogenize; and individualization in order to obliterate differences and otherness. The roots of totalitarianism are inscribed in the spatial matrix concretized by the modern nation-state – a matrix that is already present in its relations of production and in the capitalist division of labor. (107)

The production of capitalist spatiality, however, is no once-and-for-all event. The spatial matrix must constantly be reinforced and, when necessary, restructured – that is, spatiality must be socially reproduced, and this reproduction process is a continuing source of conflict and crisis.

The problematic connection of social and spatial reproduction follows straightforwardly. If spatiality is both outcome/embodiment and medium/presupposition of social relations and social structure, their material reference, then social life must be seen as both space-forming and space contingent, a producer and a product of spatiality. This two-way relationship defines – or perhaps, redefines – a socio-spatial dialectic which is simultaneously part of a spatio-temporal dialectic, a tense and contradiction filled interplay between the social production of geography and history. To provide the necessary recomposition of Marx's familiar dictum: We make our own history and geography, but not just as we please; we do not make them under circumstances chosen by ourselves, but under circumstances directly encountered, given and transmitted from the historical geographies produced in the past.

The general argument I have presented can be briefly summarized in a sequence of linked premises:

1. Spatiality is a substantiated and recognizable social product, part of a 'second nature' which incorporates as it socializes and transforms both physical and psychological spaces.

2. As a social product, spatiality is simultaneously the medium and outcome, presupposition and embodiment, of social action and relationship.

3. The spatio-temporal structuring of social life defines how social action and relationship (including class relations) are materially constituted, made concrete.

4. The constitution/concretization process is problematic, filled with contradiction and struggle (amidst much that is recursive and routinized).

5. Contradictions arise primarily from the duality of produced space as both outcome/embodiment/product and medium/presupposition/ producer of social activity.

6. Concrete spatiality – actual human geography – is thus a competitive arena for struggles over social production and reproduction, for social practices aimed either at the maintenance and reinforcement of existing spatiality or at significant restructuring and/or radical transformation.

7. The temporality of social life, from the routines and events of day-to-day activity to the longer-run making of history (*évènement* and *durée*, to use the language of Braudel), is rooted in spatial contingency in much the same way as the spatiality of social life is rooted in temporal/historical contingency.

8. The materialist interpretation of history and the materialist interpretation of geography are inseparably intertwined and theoretically concomitant, with no inherent prioritization of one over the other.

Taken together, these premises frame a materialist interpretation of spatiality that is only now taking shape and affecting empirical research. They still have to be teased out of the current literature since, for the most part, they have remained implicit. Often research which brilliantly illuminates the arguments behind these premises is not immediately recognized as such, even by the researchers themselves, as was noted in the case of Foucault. Or else, well-intentioned exemplifications end up clouding more of the arguments than they illuminate by falling back into the traps of spatial separatism.

Furthermore, powerful and persistent barriers still remain to accepting a materialist interpretation of spatiality and an assertive historical-geographical materialism aimed specifically at understanding and changing capitalist spatializations. The most rigid of these barriers arise from an unyielding Marxist, if not more generally post-Enlightenment, tradition of historicism which reduces spatiality either to the stable and unproblematic site of historical action or to the source of false consciousness, a mystification of fundamental social relations. Historicism blocks from view both the material objectivity of space as a structuring force in society and the ideational subjectivity of space as a progressively active part of collective consciousness. Stated somewhat differently, in terms that would take on new meaning in the debates of the 1980s on the restructuring of critical social theory. Western Marxism's conceptualization of the interplay between human agency and social structure has remained essentially historical, defined in the praxis of making history. Spatiality, as the praxis of creating human geography, still tends to be pushed into an epiphenomenal shadow as history's mirroring container. Historicism continues to be one of the most forceful nineteenth-century monuments that must be destroyed before critical social theory and Western Marxism can successfully envision the spatiality of contemporary social life.

Can Western Marxism be spatialized without inducing the aura of an anti-history? If so, how can the historical imagination be made to accommodate a space which is as rich and dialectical as time? These are challenging questions which have rarely, if ever, been asked before the present decade. One possible approach in response to these questions is through logical persuasion, the straightforward rational assertion of a socio-spatial dialectic, a historical and geographical materialism, a space-time structuration of social life. By now, this trajectory of theoretical assertion will be familiar to the patient reader. But it is not enough. A promising alternative path, which took me eventually to the study of urban restructuring in Los Angeles (and which will eventually take the reader there as well), is one of empirical demonstration, the application of a materialist interpretation of spatiality to contemporary 'real world' issues and politics. This turn to empirical research will undoubtedly be vital to the future development of an historical geographical materialism and a reconstructed postmodern critical social theory.

There is, however, another path, rarely followed these days, which departs from theoretical assertion in the opposite direction, looking at the 'backward linkages' rather than the empirical and political 'forward linkages' of theory formation. This is a path into the even more slippery and abstract realm of ontology, the meta-theoretical discourse which seeks to discover what the world must be like in order for knowledge and human action to be possible, what it means to *be* (Bhaskar, 1975). Assuming that there is little of importance left to discover in ontological discourse, with its characteristic distancing from praxis, most Western Marxists have been reticent to venture very far along this backward-looking path. But it is a worthwhile journey to take, for it can help us find some still-missing connections between space, time and being, and hence between the makings of history, human geography, and society.

## Back to Ontology: on the Existential Spatiality of Being

The literature of existentialism and existential phenomenology is rich in ontological discussion arising from its central concern for understanding the structures of human existence, of being, and especially of being-in-the-world. An emphasis on the active emplacement and situation of being-in-the-world is sometimes used to distinguish existential phenomenology from a 'vulgar' existentialism, entrapped in pure contemplation of the isolated individual, in what Henri Lefebvre once called the 'excremental

philosophy' of Sartre's major early work: *Being and Nothingness.*[4] But
rather than opening up the debate on the similarities and differences
between existentialism and phenomenology, I find it more fruitful to draw
upon their shared accomplishments to extract what both have to say about
the spatiality of being.

One of the most explicit treatments of existential spatiality can be
found in Martin Buber's work, 'Distance and Relation' (1957). Buber
presents spatiality as the beginning of human consciousness, 'the first
principle' of human life:

> It is the peculiarity of human life that here and here alone a being has arisen
> . from the whole endowed and entitled to detach the whole from himself as a
> world and to make it opposite to himself.

This original existential capacity to separate the individuated Human Being
from the whole of Nature, the world of things, revolves around what
Buber calls 'the primal setting at a distance'. Human beings alone are
able to objectify the world by setting themselves apart. And they do so
by creating a gap, a distance, a space. This process of objectification
defines the human situation and predicates it upon spatiality, on the
capacity for detachment made possible by distancing, by being spatial to
begin with.

It is in this sense that spatiality is present at the origin of human
consciousness for it permits – indeed it presupposes – the fundamental
existential distinction between being-in-itself (the being of non-
conscious reality, of inanimate objects, of things) and being-for-
itself, the being of the conscious human person. Objectification, the
primal setting at a distance, relates to what Sartre calls 'nothingness', the
physical cleavage between subjective consciousness and the world of
objects that is necessary for being to be differentiated in the first place,
for being to be conscious of its humanity. In this essential act, this

---

4. Lefebvre was the earliest and most cutting critic of Sartre's work. Bent on dampening the growing enthusiasm on the left for existentialism in the 1940s, he attacked *Being
and Nothingness* (*L'Etre et le néant,* 1943, English translation by Hazel Barnes, 1956) as a
collection of discarded and reactionary ideas, ideas which he himself had toyed with in his
youth but had abandoned as peripheral to a non-reductionist development of Marxism. See
*L'Existentialisme* (1946) for Lefebvre's critique, and Mark Poster, *Existential Marxism in
Post-War France* (1975) for a critical view of the Sartre-Lefebvre confrontation. Later in
his life, Sartre would agree almost entirely with Lefebvre's critique of *Being and Nothingness.* He writes that '*L'Etre et le néant* traced an interior experience, without relating it to
the exterior experience ... of the petty-bourgeois intellectual that I was ... Thus, in *L'Etre
et le néant,* what you could call "subjectivity" is not what it would be for me today: the
*décalage* [shifting point?] in a process by which an interiorization re-exteriorizes itself in an
act' (*Situations, IX,* 102; cited in Fell, 1979, 478, footnote 59. Fell translates *décalage* as
'small margin').

original spatialization, human consciousness is born (although borne may be just as appropriate). Nothingness is thus nothing less than primal distance, the first created space, the vital separation which provides the ontological basis for distinguishing subject and object.

Objectification, detachment, and distancing, however, are but one existential dimension of consciousness, the basis for only a minimal definition of being. To be human is not only to create distances but to attempt to cross them, to transform primal distance through intentionality, emotion, involvement, attachment. Human spatiality is thus more than the product of our capacity to separate ourselves from the world, from a pristine Nature, to contemplate its distant plenitude and our separateness. In what may be the most basic dialectic in human existence, the primal setting at a distance is meaningless (one of existentialism's most important concepts) without its negation: the creation of meaning through relations with the world. Thus, as Buber argues, human consciousness arises from the interplay – dare I add unity and opposition? – of distancing and relation. Entering into relations, being-in-the-world, Heidegger's *Dasein*, Sartre's *L'Etre pour-soi* or *être-là* ('being-there'), is not possible without distancing, without the ability which allows us to assume a point of view of the world. But in this ability is also a will to relate, a necessary impulsion to overcome detachment, as the only means whereby we can confirm our existence in the world, can overcome meaninglessness and establish identity. In this way, Buber reasons, the two movements 'contend with one another, each seeing in the other the obstacles to its own realization'. Subjectivity and objectivity thus reconnect in a dialectical tension that gives place to being, that produces a milieu, a humanized second-nature.

In this reconnection of subject and object lies a dilemma that encapsulates the core of the existential critique, the dilemma of meaning versus alienation. Objectification is part of the human condition, but setting oneself apart from the world is also the source of alienation. Entering into relations with the world, the creative connection between the human subject and the objects of his/her concern, is a search to overcome alienation, yet this too threatens to be alienating when it reduces the subjective self, when the subject is objectified through relations with the world. Thus existential alienation is a state of separation both from oneself and from the objective world – from the very means and meaning of existence.

In the classic works of existentialism and phenomenology, this dialectical tension between the reality of alienation and the need to overcome it tends to be rooted in time, in the temporality of *becoming*, and consequently in 'biography formation' and the making of history. But it is simultaneously and intrinsically spatial, a point which Sartre, Heidegger, and Husserl, perhaps the three most prominent and influential existential phenomenologists of the past century, emphasize in their major works. Theirs is an explicitly situated (or to use Husserl's term, 'regional' – see Pickles, 1985) ontology in which existence and spatiality are combined through intentional and creative acts inherent to being-in-the-world, entering into relations, involvement. This existential spatiality gives to being a place, a positioning within the 'lifeworld' (Husserl's *Lebenswelt*). This *emplacement* is a passionate process that links subject and object, Human Being and Nature, the individual and the environment, human geography and human history. Sartre, in *Search for a Method*, provides a particularly apropos intervention:

> [T]he 'milieu' of our life, with its institutions, its monuments, its instruments, its cultural 'infinites' ... its fetishes, its social temporality and its 'hodological' space – this *also* must be made the object of our study.... A product of his product, fashioned by his work and by the social conditions of production, man *at the same time* exists in the milieu of his products and furnishes the substance of the 'collectives' which consume him. At each phase of life a short circuit is set up, a horizontal experience which contributes to change him upon the basis of the material conditions from which he has sprung ...
>
> The aim then is to construct horizontal syntheses in which the objects considered will develop freely their own structures and their laws. In relation to the vertical synthesis, this transversal totalization affirms both its dependence and its relative autonomy. By itself it is neither sufficient nor inconsistent.[5]

Place thus comes to being from the 'short circuits' inherent in the horizontal experience, the push and pull of spatiality on the vertical–temporal trajectory of life. This emplacement grounds being in place, but, alas, also becomes the source of still another ontological dilemma, one which Sartre, Heidegger, Husserl, and many others struggled with,

---

5. Sartre, 1968, 79–80. *Search for a Method* is a translation (by Hazel E. Barnes) of *Question de méthode*, an essay which Sartre wanted originally to put at the end of his massive *Critique de la raison dialectique* (1960) but eventually published as its introduction. The term 'hodological' space, Barnes notes, is derived from Kurt Lewin and refers to the environment viewed in terms of our personal orientation. The references to 'horizontal experience' and 'horizontal syntheses', however, are clearly derived from Henri Lefebvre, who is uncharacteristically praised by Sartre in a long footnote (1968 trans., 51–52). In his Preface to the combined publication – the title page reads *Critique de la raison dialectique* (*précedé de Question de méthode*) – Sartre also notes that *Search for a Method* derived from an invitation by a Polish journal to contribute an essay on existentialism to a special issue on French culture. Sartre was annoyed by this constraint on his desire to write on 'the present contradictions of philosophy' especially in French Marxism, but someone else had been asked to cover these themes: Henri Lefebvre. Poster (1975) examines in some detail Lefebvre's critique of *Being and Nothingness*, but I have not been able to find any treatment of the subsequent interconnections between these two leading French philosophers.

especially in their more mature writings. The dilemma connects back to what has been called the problem of 'divided being', the powerful separation between the thinking subject and the 'grounding' object, the transcendental ego and the world as lived, a problem that has polarized the modern ontological tradition at least as far back as the Cartesian opposition of *res cogitans* and *res extensa*. The original emplacement, the source of the problem, is viewed in much the same way by Sartre and Heidegger. The former describes 'human reality' as the being which causes place to come to objects: 'It is not possible for me not to have a place'. For the latter, *Dasein* 'names that which should first be experienced, and then properly thought of as Place – that is, the locale of the truth of Being' (see Fell, 1979, 31). But once being takes place, how is the relation between place and being to be understood? As separate spheres? As interdependencies? As shaped entirely by the forcefulness of the absolute ego? As shaped entirely by the materiality of place? I suggest that these are the ontological interrogations from which all social theory springs.

Sartre and Heidegger each try to overcome the dualization of this emplacement through a fusion which appears to resemble the assertion of a socialized second nature emerging from the groundedness of being, a 'third way' through the double bind. Joseph Fell, in *Heidegger and Sartre: An Essay on Being and Place*, describes the Heideggerian third way:

> The originary unity is a togetherness in apartness and an apartness in togetherness. Earth and thought are certainly distinct, yet they belong together and from the beginning play into each other as *Intelligible Earth.* (1979, 385, author's emphasis)

Fell goes on to compare this dialectical fusion (my words) to the classical opposition between the universal and the particular, which must also be seen as fused, as 'the same' to use his words. This endows contact with the lifeworld with the significant philosophical status that it has traditionally lacked, especially given the tendency of conventional ontologies to assert the ego as absolute and supreme. But after giving us this intelligible lifeworld as an identifiable spatialization of being, Heidegger proceeds to envelop it in temporality. In *Being and Time* (1962, a translation of *Sein und Zeit*, 1927), he writes:

> Spatiality seems to make up another basic attribute of Dasein corresponding to temporality. Thus with Dasein's spatiality, existential–temporal analysis seems to come to a limit [Sartre's 'short circuit'?], so that this entity which we call "Dasein" must be considered as 'temporal' 'and also' as spatial coordinately. Has our existential–temporal analysis of Dasein thus been brought to a

halt by that phenomenon with which we have become acquainted as the spatiality that is characteristic of Dasein, and which we have pointed out as belonging to Being-in-the-world? (418)

Here we are brought to the ontological edge of spatio-temporal dialectic, an existential tension between the horizontal and vertical experience, the possibility of a balanced interpretation of space, time, and being. The primacy of time, however, cannot be resisted. Heidegger answers his question this way:

> Dasein's constitution and its ways to be are possible ontologically only on the basis of temporality, regardless of whether this entity occurs 'in time' or not. Hence Dasein's specific spatiality must be grounded in temporality.

Heidegger re-opens the question of the spatio-temporality of being in his later works and softens his temporalization enough to assert the importance of place as a fundamental ontological category 'concealed in history'. But his construction of a more fully spatialized ontology is never completed. Heidegger's half-hearted glorification of place was eventually used to feed the flames of Nazism, at times with his own apparent encouragement, and would contribute to the personal and philosophical isolation that marked his being-in-the-(post-war) world until his death in 1976, alone in his birthplace in the Black Forest.

Whereas Heidegger's historicization of space was ultimately conservative, disengaging, and spiritual, Sartre's spatio-temporal path was much more radical, activist, and materialist, leading him smack into Marxism via his 'search for a method'. Sartre's intrepretation of history and the intelligibility of the lifeworld is couched in praxis, which in *Critique of Dialectical Reason* (1976) he links to a movement whose fundamental direction is determined by 'scarcity' and which provokes the formation of groups to struggle collectively for such scarce necessities, such 'worked matter'. Sartre describes this horizontalized vertical movement as a spiral. Fell explains:

> [A spiral] represents the circle stretched out in three dimensions – an exploded 'center' that has to keep on circling, optimally at an ever-higher level, in an effort to reach the ideal future time at which the organism will, as at the beginning, again be in equilibrium or in a stable relation with its environment. Its vertical axis represents the linearity of man's historical quest for unification. Its horizontal axes (each revolution) represents the 'deviations' from strict linearity necessary along the way, the repeated externalizations and internalizations in which man both makes and is made by a material environment that both threatens and supports his project. Thus the inward-and-outward horizontal movement expresses a dialectical mediation.... [The spiral] represents, then, materialization of the existentialist project or the

insertion of Sartrean 'revolutionary consciousness' into a material history as the price of its efficacy. (Fell, 1979, 348–49)

Here again we are at the edge of a philosophy of historical geography, of space–time structuration, of a balanced and unprioritized ontology and epistemology of space, time, and being. Yet Sartre too is never quite able to escape his ontological historicism, the explicitly privileged 'first' place he (and his historical materialism) gives to the making of history. Rather than applying his *Critique* to a historical geography, such as might have appeared in his unfinished work on the Soviet Union, Sartre chose to devote most of his later years to biography, that most flagrant vehicle of the individualized historical imagination.[6]

Despite their ultimate conclusions, in their lives and in their thoughts, Heidegger and, especially, Sartre contribute significantly to the reassertion of space in social theory and philosophy. In the wake of the ontological struggles that punctuated the achievements of these two twentieth–century existential phenomenologists has come a reawakening to the spatiality of being, consciousness, and action, a growing awareness of the possibility of spatial praxis, an increasingly recognized need to rethink social theory so as to incorporate more centrally the fundamental spatiality of social life.

---

6. Sartre's 'totalizing' three-volume biography of Gustave Flaubert, *L'Idiot de la famille*, was published in 1971 and 1972. For an interesting contemporary attempt to give biography formation an explicit time–*geography*, see Pred (1984, 1986).

# 6

# Spatializations: a Critique of the Giddensian Version

[It] is ... as if each new aggression from the cosmic exterior appeared at the same time as a disparity to be absorbed and as the perhaps unique opportunity to recommence, on new grounds, the great totality-concocting which tries to assimilate ancient and indestructible contradictions, that is, to surpass them in a unity which is at long last rigorous – a unity which would be manifested as a cosmic determination.... One may envisage the circular movement in a three-dimensional space, as a spiral whose many centers are ceaselessly deviated and ceaselessly rise by executing an indefinite number of revolutions around their starting point. Such is the personalizing evolution, at least up to the moment ... of sclerosis or regressive involution. In this latter circumstance the movement indefinitely repeats itself by passing the same places again or else is an abrupt fall from a higher revolution to some inferior revolution. (Sartre, *L'Idiot de la famille*, I, 1971, 656–57, trans. in Fell, 1979, 348)

## Redoubling the Helix: Space–Time and Anthony Giddens

For more than a decade Anthony Giddens has been spiralling toward a critical reconceptualization of social theory in a remarkably linked sequence of books which have established him as one of the foremost contemporary interpreters of social theory writing in English. From his first critical reviews of the origins of sociology to his most recent theoretical syntheses, Giddens's project has evolved in the form of a helix. His arguments move persuasively forward through the accumulated antinomies that have traditionally divided social science and philosophy, but always curve back again to gain new perspective on the historical roots of sociological theory and analysis. This distinctive trajectory and style were set in his earliest works, where he attempted to recast social theory around a syncretic and critical appropriation and modernization of the classical theoretical programmes of Durkheim, Weber and Marx. With each new advance in his thinking, Giddens almost dutifully returns to evoke and reconsider this continental European inheritance from a different vantage point, somewhat more distant, but never so far as to lose sight of the enduring traditions.

In *New Rules of Sociological Method* (1976), for example, Giddens condensed his evolving critique around an analytical theory of meaning and action built upon a constructive reevaluation of interpretive sociology and hermeneutics. The helix path cut through broad realms of twentieth-century humanisms and action philosophies to centre on the creative force of human agency and praxis. It then curved back again to excoriate persistent functionalism (a recurrent theme in Giddens's work), resift through the Durkheimian legacy, and exorcise once more the ghost of Talcott Parsons, whose enervating theory of action so powerfully shaped post-war academic sociology and lingers in the background of most of Giddens's writings.

In *Central Problems in Social Theory* (1979) an important shift occurred. Giddens engaged his invigorated action theory with a sympathetic critique of the main currents of structuralist thought. Through this inflammatory conjunction of human agency and determinative structure, Giddens drew together two theoretical discourses which had developed through the twentieth century in explosive and unreconciled opposition. In *Central Problems* the dialectical engagement of agency and structure, subjectivity and objectivity, was assertively placed at the core of social theory, reconceptualized by Giddens in a budding theory of structuration which situated praxis and social reproduction in '*time and space* as a continuous flow of conduct' (1979, 2). This comprehensive confluence of ideas marked, for Giddens, the culmination of one spiral of critical reinterpretation and the beginning of another, more committed and constructive than the first.

Each of Giddens's books contains the seeds of its sequel, a pattern never more evident than in the link between *Central Problems* and his next major work, *A Contemporary Critique of Historical Materialism* (Volume I, 1981). *Critique* is much more than – and less than – an effective reinterpretation of Marx's historical materialism, an inching forward to glance back again to the nineteenth century. Although Marx, Durkheim, and Weber continue to fill more index space than any other authors, *Critique* became Giddens's most explicit and committed assertion of his own conceptualization of social theory, a constructive affirmation of the theory-generating capacity of the agency-structure nexus. It is cautiously offered as a propaedeutic, 'a stimulus to further reflection rather than ... approaching an exhaustive analysis of the major issue it raises' (1981, 24). Propaedeutic or not, *Critique* is Giddens's most original and therefore most vulnerable book, at once a cause for celebration and an invitation to critical reappraisal of the author's entire theoretical project.

*Critique* must be evaluated at both a substantive and a theoretical level, and as simultaneously a deconstructive critique and an attempted reconstructive affirmation. Giddens previews his approach to historical materialism in *Central Problems* (1979, 53), where he states that 'Marx's writings still represent the most significant fund of ideas that can be drawn upon in seeking to illuminate problems of agency and structure'. Their powers of illumination, however, must be brightened by selectively discarding an encumbrance of 'mistaken, ambiguous or inconsistent' analytical concepts and the many errors of subsequent Marxisms. Stripping away this encumbrance is the titular objective of *Critique.*

Many of the targets selected by Giddens are familiar themes of discussion within the contemporary Marxist literature: the inadequacy of Marx's evolutionary schema and outdated anthropology; the dangers of economism and structuralist determinism; the overuse of functionalist categories and explanation; the absence of appropriate theories of the state, of politics, of urbanization, of power. There is an attack on the mode of production as an analytical concept, a denial of the incessantly progressive augmentation of productive forces, a refusal to accept 'all history' as the history of class struggle. The phalanx of critical dismissals will no doubt anger and annoy some Marxist readers. Others will argue, with merit, that precisely the same issues have been addressed more effectively by critical theorists less averse to accepting the label 'Marxist' than is Giddens.

Yet, despite his grumblings, Giddens remains peculiarly accepting and sympathetic, committed to the centrality of historical materialism in the construction of critical social theory. Indeed, the critique of historical materialism he offers is primarily an accessory to the application and elaboration of Giddens's theory of structuration and, in particular, the embedded distinction between 'class-divided' and 'class' society posited in *Central Problems.* The substantive chapters of *Critique* revolve around this distinction in an attempt to address the specificity of industrial capitalism in comparison with prior phases in world history. The differences between class-divided societies (primarily agrarian states in which classes exist, but for which 'class analysis does not serve as a basis for identifying the basic structural principle of organization' – 1981, 7) and class society (that is, capitalism, wherein class-conflict, struggle, and analysis are essential and central) unfold in a series of critical essays which are stuffed with 'preliminary learning', loosely synthesized propaedeutic insights which I suspect would not easily withstand rigorous critical analysis, especially perhaps by Giddens himself.

Chapter 3, 'Society as time-traveller: capitalism and world history', is an analysis of the contradictions between Marx's evolutionary schema and the more guarded insights contained in the *Formen* section of *Grundrisse*. This is followed by 'Time-space distanciation and the generation of power' (an assertion of the importance of time-space relations versus relations with nature in a significantly reoriented materialist interpretation of history); 'Property and class society' (on the generation of class society in the interlocking of capital and wage-labour in a 'dialectic of control' shaped by the private ownership of property); 'Time, labour and the city' (on the commodification of time and space in everyday life under capitalism, an eclectic synthesis of Lefebvre, Castells, Harvey, Mumford, Wirth, Christaller, Sjoberg, et al.); 'Capitalism: integration, surveillance and class power' (a further exploration of the specificity of capitalism in terms of means of control, the role of the state, and the emergence of world systems of intersocietal integration); 'The nation-state, nationalism and capitalist development' (an interesting excursion from Montesquieu to the new international division of labour); and 'The state: class conflict and political order' (a creative, but limited, tour of the current debates on the theory of the state). *Critique* ends, characteristically, with the seeds of its projected sequels (*The Nation State and Violence* and *Between Capitalism and Socialism*), enmeshed in a discussion of 'Contradiction and exploitation'.

Before allowing Giddens to jump ahead to another stage in his helix path, however, some careful consideration must be given to the conceptualizing arguments which frame these substantive chapters and are presented in *Critique* (1981, 3) as 'elements of an alternative interpretation of history'. In particular, the *theory of structuration* must be submitted to the same 'positive critique' that Giddens has so successfully applied to others. In doing so, it can be argued that the spiralling trajectory which has marked Giddens's long project and propelled him into the perspicacious achievements of *Critique* may have become its own conceptual trap, constraining further theoretical development rather than generating it. A propaedeutic book perhaps deserves a propaedeutic evaluation, an invitation to further reflection rather than an exhaustive analysis.

Giddens's theory of structuration builds upon and elaborates Marx's pithy maxim that 'men make history, but not in circumstances of their own choosing', still the most evocative encapsulation of the agency–structure relation in social theory. To the making of history, Giddens adds, awkwardly at first and without full awareness of its implications, what can be described as the 'making of geography', the social production of space embedded in the same dialectic of praxis. *Critique* calls for the injection of temporality and spatiality into the core of social theory, and binds and brackets the theory of structuration in time–space relations. 'All social interaction', Giddens writes (1981, 19), 'consists of social practices, situated in time–space, and organized in a skilled and knowledgeable fashion by human agents'. Knowledgeability and action, however, are always 'bounded' by the structural properties of social systems, which are simultaneously the *medium* and *outcome* of social acts (forming what Giddens calls the 'duality of structure'). Social systems are thus conceived as *situated practices*, patterned (structurated) relationships socially reproduced across time and space, as history and geography.[1]

The theory of structuration is amplified through a combination of three discourses which serve to link the articulation of space–time relations directly to the generation of power and the reproduction of structures of domination. Heidegger's philosophy of Time and Being, Althusser's structuralist schemata, and the writings of modern geographers on such concepts as 'time–geography' and the subjectivity of distance, are recomposed by Giddens to describe 'how form occurs', how situated practices conjoin 'moments' temporally, structurally, and spatially in the constitution of social life. What comes through most clearly in the cloud of neologisms and revamped vocabulary (for which Giddens understandably begs indulgence) is an institutional emphasis on the operation of power, within which Giddens posits another definitive bifurcation. Power and domination are coupled in the structuration of allocative control (over the material world) and authoritative control (over the social world). Allocation and authority thus come to define, respectively, the realms of the economic and the political, and they connect the general theory of structuration to the themes and literature referred to in the subtitle of *Critique: Power, Property and the State*.

The theory of structuration outlined in *Critique* remains elusive, however, and much more appealing in intent than in execution. Part of the problem lies in the immensity of the task and in the disparate languages being unconventionally conjoined around the agency–structure linkage. In addition, Giddens's recurrent strategy in formulating theoretical arguments has been to spin off interlocking classificatory schema, a practice which becomes intractably dense in *Critique*, too often confusing rather than clarifying the argument. More fundamentally, however, the theory of structuration is built around a gener-

---

1. It is interesting to note that Giddens consistently emphasizes the combination 'time–space' but never explicitly uses 'historical geography'.

ative premiss which requires a more formidable adjustment in theoretical perspective than Giddens is able to achieve.[2] Although his repeated intention is to project both temporality and spatiality into the heart of critical social theory, presumably in the explicit balance of time–space, Giddens – very much like Heidegger – manages unintentionally to perpetuate the long-standing submergence of the spatial under the ontological and epistemological primacy of time and history. For Giddens, history and sociology become 'methodologically indistinguishable', but the analysis of spatial structuration remains peripheral, an insightful accessory.

Giddens's discovery of the 'writings of modern geographers' and the spatiality of structuration is, nevertheless, the most important new ingredient both in *Central Problems* and in *Critique*. It distinguishes these works more propitiously than anything else from all the author's earlier contributions, in which the spatiality of social life remained virtually invisible. Unfortunately, the growing contemporary debate on social theory and spatial structure, on the dialectics of society and spatiality, is barely seen by Giddens, who presents his discovery almost as if he were a lonesome pioneer. This leads him to draw upon disjointed pieces of the writings of such key contributors to this debate as Lefebvre, Foucault, Harvey, Castells, and Poulantzas, without recognizing that they have been providing the theoretical substance for an alternative conceptualization of the time–space constitution of social systems so central to *Critique*. In *State, Power, Socialism* (1978), for example, Poulantzas refocused his analysis of the institutional materiality of the state around the formation and transformation of 'spatial and temporal matrices', manifested in the themes of territory and tradition. As noted in the previous chapter, these matrices were defined as the 'presuppositions' (versus merely preconditions or outcomes) of capitalism, implied in the relations of production and the division of labour. Temporality and spatiality are presented together as the concretization of social relations and social practice, the 'real substratum' of mythical, religious, philosophical, and experiential representations of space–time. *Critique* would have been much richer had Giddens incorporated the explicitness and balance of Poulantzas's interpretation, both at the level of theory and in the substantive chapters on the state and nationalism, where their absence is most disturbing. Giddens's exposition of time–space distanciation, presencing and absencing, the commodification of time and space, allocation and authorization, would also have become clearer and more

---

2. Given the discussion in the previous chapter, it might be more accurate to describe this generative premise as an *ontological* assertion derived primarily from Heidegger, whose works have been particularly influential in Giddens's theorizations.

comprehensible. Instead, no mention is made of this crucial dimension of Poulantzas's last major work.

The irony of *Critique* is that Giddens misses what his helix path has so productively achieved over the past decade: an opportunity to re-evaluate and reconstitute the classical contributions of Marx, Weber, Durkheim and the twentieth–century achievements of hermeneutics and structuralism. There is another helix of critical theory still to be written that would trace the history (and geography?) of the theoretical primacy of time over space to its generative roots. In this spiral, Durkheim, Weber, and Marx are again primary sources. It was in the anti-Hegelian wellsprings of historical materialism that revolutionary time and history displaced spatiality (in the spiritual form of the Hegelian state and terri-torial consciousness) and relegated it to the status of idealistic and diver-sionary fetishism. The development of an effective materialist theory of the state, of nationalism and regionalism, of the territorial collectivity and consciousness, has been constrained ever since. Similarly, the theor-etical programmes of Durkheim and Weber, building a relatively space-less social science based on differing interpretations of the link between individual action and collective consciousness, also peripheralized the spatial into an almost mechanical externality. Spatiality became a passive mirror/container to the forceful play of human agency and social process set free from 'environmental' determination.

Hermeneutics and structuralism reproduced much of this traditional imbalance. Existential phenomenology, despite the inherently spatial quality of such concepts as *Dasein, Etre-là*, Being-*there*, continued to concentrate on the temporality of Being and Becoming. For Heidegger in particular, the space of being remained a chronic problem, in more ways than one. Structuralism's celebration of the synchronic, in compar-ison, was filled with promising spatial metaphors but relatively little explicit spatial analysis. Nevertheless, hermeneutics and structuralism both opened new windows through which to re-engage time–space relations in a more appropriate symmetry.

Persistently combative and procrustean as structuralism and hermeneu-tics have been, their recent and still tentative conjunction around the agency–structure relation (of which Giddens's work is but one major example) has demanded an appropriately dialectical nexus, with no enforced priority of agency over structure or the reverse. Significantly, this dialectical connection of agency and structure has been accompanied by increasing attention to another traditional duality, the spatial and the tem-poral, which calls for a similar conceptualization: epistemologically co-equal, dialectically related in their material expression, unified in praxis, and positioned at the very heart of critical social theorization.

Giddens edges close to this critical reconceptualization, certainly closer than any other contemporary sociologist writing in English. His theoretical 'space', however, remains too constrained. There is no mention in *Critique*, for example, of his Cambridge co-resident, Derek Gregory, whose work on social theory and spatial structure in the context of the agency–determination relation has so brightly illuminated the contemporary geographical literature.[3] There is also a too narrow and blinkered appropriation of French social theory. In particular, the extensive works of Lefebvre on the spatiality of social life and social reproduction, on the dialectic of agency and structure embedded in the social production of space, cannot be reduced to his commentaries on *le quotidien* and an errant reification of the 'urban', as Giddens does (following, as too many others have done, the voice of Castells in *The Urban Question*).

Although these weaknesses might be defined as 'structural', they are not, of course, conclusively determined, especially given the reflective and knowledgeable human agent involved. Soon after the publication of *Critique* and before the completion of its promised sequels, Giddens moved on to another level of theoretical development, passing the same places again but with greater clarity and a more formalizing intent. In *The Constitution of Society* (1984), Giddens simultaneously responded to his critics, laid bare the eclectic sources of his recent 'personalizing evolution', and carefully consolidated a totality-concocting theory of structuration. The propaedeutic seeds of *Critique* have now blossomed into a mature and orderly garden, with each flowering species carefully labelled as to its ontogenetic and phylogenetic heritage. *Constitution* thus offers another opportunity to consider the trajectory advanced by *Critique* and to reconstruct on firmer grounds the Giddensian version of the reassertion of space.

## The Constitution of Society and the Reconstitution of Social Theory

In an interview with Derek Gregory in *Society and Space*, Giddens described his distinctive personal project:

> I don't think of myself as working in any innovative way on epistemological issues, and I try to 'bracket' them to some substantial degree. What I am trying to do is to work on essentially what I describe as an *ontology* of human

---

3. See Gregory (1978) in particular. After the publication of *Critique* there is much more contact between Giddens and Gregory. See Gregory (1984), Gregory and Urry (1985), and Gregory's entry on 'structuration theory' in *The Dictionary of Human Geography* (Johnston et al., 1986).

society, that is, concentrating on issues of how to theorise human agency, what
the implications of that theorising are for analysing social institutions, and
then what the relationship is between those two concepts elaborated in
conjunction with one another.... I don't think it either necessary or possible
to suppose you could formulate a fully-fledged epistemology and then somehow securely issue out to study the world. So my idea is to fire salvoes into
social reality, as it were; conceptual salvoes, which don't provide an overall
consolidated epistemology. (Gregory, 1984, 124)

What emanates from these conceptual salvoes in *Constitution* is a
reformulated theory of being, of the nature of social existence. Placed in
proper perspective, *Constitution* stands out as the most rigorous,
balanced, and systematic ontological statement currently available on
the spatio-temporal structuration of social life. Its position and lineage
within the discourse of critical social theory are obvious, but its accomplishment extends more broadly, via the philosophical tracks laid down
by the efforts of Husserl, Heidegger, and Sartre, to give 'place' to being.
This is where its primary achievements need to be located.

The intentional absence of a formal epistemology makes any simple
and direct translation of Giddens's ontology into demonstrative empirical research rather difficult, while its necessary conceptual inventiveness
continues to provoke misunderstanding, especially amongst those who
seek such direct and simple empirical insight from Giddens's work. *The
Constitution of Society* nevertheless provides illuminating, if complexly
sinuous, guidelines to empirical analysis and, in particular, to a critical
reinterpretation of the historical geography of capitalism. It does not
present easy formulas and blueprints, nor does it propound rigidly categorical stances on the theoretical paths to be followed. But this is its
strength, not its weakness.

The structuration theory in *Constitution* is an elastic synthesis of the
almost endless concatenation of associated dualisms that has followed
upon the too often frozen opposition of subjectivity and objectivity.
Agency and structure, the individual and the societal, are flexibly
combined by Giddens and this ontological flexibility and fusion is the
primary message.

The key synthesizing concepts asserting this ontological balance can
be lifted from the conveniently appended Glossary (pages 373–7). Our
inherited concept language is so distorted with regard to spatio-temporal
relations that it must be radically restructured to express the articulation
of space, time and social being, a task which Giddens self-consciously
takes on in *Constitution*. The resulting conceptual glossary is an artful
balancing act that consistently inserts space adjoined with time, but
never space alone, into the constitution of society.

*Contextuality:* The situated character of interaction in time–space, involving the setting of interaction, actors co-present and communication between them

*Locale:* A physical region involved as part of the setting of interaction, having definite boundaries which help to concentrate interaction in one way or another

*Regionalization:* The temporal, spatial or time–space differentiation of regions either within or between locales; regionalization is an important notion in counter-balancing the assumption that societies are always homogeneous, unified systems

*Social integration:* Reciprocity of practices between actors in circumstances of co-presence, understood as continuities in and disjunctions of encounters

*System integration:* Reciprocity between actors or collectivities across extended time–space, outside conditions of co-presence

*Time–space distanciation:* The stretching of social systems across extended time–space, on the basis of mechanisms of social and system integration

Despite the conceptual advances, the temporal chaperon becomes too protective on occasion, for Giddens is determined to acknowledge space without succumbing to the disciplinary biases of Modern Geography and their peculiar separatisms. There is much less caution, however, with regard to history and its disciplinary inclinations. As a result, time and history often stand alone in *The Constitution of Society*, authoritative and allocative, much more 'established' than the less familiar geographical 'outsider'.[4] The enforced order is always 'time–space' connected in the same dominating to dominated sequence as 'core–periphery'.

Giddens thus fails again to initiate the necessary critique of historicism that must accompany the contemporary restructuring of critical social theory.[5] Nevertheless Giddens's reformulated conceptual vocabulary can be effectively appropriated to reconstruct the substance and meaning of spatio-temporal structuration. With some adaptive extension, the framework of concepts establishes a provocative social ontology

---

4. Giddens's glossary includes an entry for *historicity*: 'The identification of history as progressive change, coupled with the cognitive utilization of such identification in order to further that change. Historicity involves a particular view of what "history" is, which means using knowledge of history in order to change it.' There is no equivalent entry for *spatiality*.

5. But so too, it must be added, has every one of the spatializing social theorists I have discussed, from Foucault and Lefebvre (who come the closest) to Harvey, Mandel, and Jameson.

conducive to the development of historico-geographical materialism, one that is far better suited to the task than any other that has emerged from the encounter between Modern Geography and Western Marxism.

Giddens, to be more specific, comes closer than any other influential social theorist to uncovering what, in my view, is the most fundamental contextual generalization about the spatiality of social life: that the intelligible lifeworld of being is always and everywhere comprised of a multi-layered system of socially created nodal regions, a configuration of differentiated and hierarchically organized locales. The specific forms and functions of this existential spatial structure vary significantly over time and place, but once being is situated in-the-world the world it is in becomes social within a spatial matrix of nested locales. The topological structure is mutable and permutable, but it is always there to envelop and comprise, to situate and constitute all human action, to concretize the making of both history and geography.

Geographers and sociologists have peered at pieces of this existential spatialization and produced an impressive literature describing the particularities and hypothesized geometries of its real or expected empirical appearances.[6] The generative sources of the spatial matrix, however, have been both elusive and illusive. The failure of geography and sociology to recompose an appropriate ontology in which, to use a currently fashionable phrase, 'space matters' (rather than just being there) has kept the existential meaning of the spatial context hidden. Let us look more closely at how Giddens's approximation might be effectively extended to bring out more clearly the spatial generality and specificity of social being.

First there is the evocative concept of locale, a bounded region which concentrates action and brings together in social life the unique and particular as well as the general and nomothetic. As Giddens notes, it is a notion somewhat akin to 'place' as used in the writings of cultural geographers (where, I might add, it is often asserted as a favoured alternative to 'space' and 'region'). But it provokes even closer comparison to the use of 'place' in the ontologies of Heidegger and Sartre. For Giddens, locales refer to 'the use of space to provide the *settings* of interaction, the settings of interaction in turn being essential to specifying its *contextuality*' (1984, 118). These settings may be a room in a

---

6. Central place theory, for example, describes an idealized geometry of the spatial matrix under conditions in which market relations and distance minimizing behaviour with regard to the provisioning of social services are assumed to dominate the social production of space. Its models occasionally bear some fortuitous resemblance to the actual geographical landscapes of capitalist societies, largely because they too are structured around an assumed spatial matrix of nested locales. They represent one of the very few attempts in the history of social theory to address selected aspects of this existential spatialization.

house, a street corner, the shop floor of a factory, a prison, an asylum, a hospital, a definable neighbourhood/town/city/region, the territorially demarcated areas occupied by nation-states, indeed the occupied earth as a whole. Locales are nested at many different scales and this multi-layered hierarchy of locales is recognizable as both a social construct and a vital part of being-in-the-world.[7]

The concentration of interaction in locales is linked to another contextual specificity of social being which Giddens is hesitant to acknowledge. It might best be described as the nodality of social life, the socio-spatial clustering or agglomeration of activities around identifiable geographical centres or nodes. Nodality and centering in turn presuppose a social condition of peripheralness: for every centre there is a more or less boundable hinterland defined by a geographical dimunition in nodality that is brought about mainly through controls over access to the advantages of agglomeration. Nodality and peripheralness exist to some degree in every locale, if only as a product of individual and collective efforts to contend with the ontologically given friction of distance imme-diately imposed on being-in-the-world. Existence, the very presence of being, means having to deal with the friction of distance whether it be on the level of the 'primal setting' or in the dull routines of every-day life. A distance-ordered space–time patterning thus pervades the existential setting of human interaction and cannot be ignored in theory construction.

But neither should the friction of distance be ripped out of its social contextuality and modelled as a quasi-Newtonian 'independent variable' determining the nodality of locales, as has so often been the case in quantitative or 'scientific' modes of Modern Geography. As Giddens implies in his too brief discussion of centre–periphery distinctions and 'uneven development', the operation of allocative and authoritative power regulates the formation of centres and peripheries across the whole range of locale-settings. Trying to avoid the obscurant tactics of spatial separatism, with its inherent depoliticization of spatiality, Giddens embeds nodality and its spatial extensions in the temporality of power relations, in an axis of antecedent 'establishment' of control over people and resources that subsequently defines the state of being 'outside'. This temporal axis of differentiation intersects with that between central and peripheral regions to form the baselines for Giddens's notions of time–space distanciation and regionalization, how

---

7. Scale and hierarchy must also be seen as social constructs, not simply as existential givens. For some recent discussions of the distinctive spatial scales associated with capitalist development (at the level of the global, the nation-state, and the urban), see Taylor (1981) and Smith (1984). These works, however, are little more than initial probes into a very complex and understudied subject.

human interaction is 'stretched' over time and space in a series of unevenly developed and differentiated settings.[8] Put more simply, the spatiality and temporality of locales are contextually intertwined and inseparably connected to relations of power from outset to outcome. Central and peripheral regions are thus homologous with the creation of a primordial social opposition between the in and out of power, to hark back to my earlier argument on the nature and necessity of geographic- ally uneven development and the relations between spatiality and class (see Chapter 4).

Nodality, regionalization, and power are also involved in another contextualizing feature of social being, the creation of bounded enclo- sures which demarcate what Giddens terms the 'presence availability' (presence/absence) of human interaction. Here two additional and closely related terms, 'territoriality' and 'regionalism', need to be included in the Giddensian glossary and woven into the theory of struc- turation. Both work, in many different ways, to segregate and compart- mentalize human interaction by controlling presence/absence and inclusion/exclusion. Like the centre–periphery distinction, with which they are closely related, territoriality and regionalism express the allo- cative and authoritative power that operates in locales. To borrow from Foucault, they are products of the instrumentality of space/power/ knowledge and provide the basis for making the operation of power both spatial and temporal.

Territoriality is the more general term and contains hints of such particularized notions as sovereignty, property, discipline, surveillance, and jurisdiction.[9] It refers to the production and reproduction of spatial enclosures that not only concentrate interaction (a feature of all locales) but also intensify and enforce its boundedness. Territoriality, almost by definition, is present in every locale at least at the outer boundary (where the absence of interaction begins). But this bounding can be more or less rigid or permeable and can change shape over time. It can also exist within the locale setting. This intra-locale territoriality may or may not coincide with central and peripheral regions but it is always associated with regionalization, with spatio-temporal divisions of activity and relation. Regional differentiation within and between locales is in

---

8. Giddens presents a simple diagram (1984, 131) to describe these relations. The vertical axis has 'established' on top, 'outsiders' at the bottom. It is crossed by a horizontal axis running from 'central regions' to 'peripheral regions'.

9. I began to explore the concept of human territoriality and its relation to the political organization of space in the late 1960s (see Soja, 1971). Much of this work had to be purely defensive, for the then prevailing view of territoriality was filled with bio-ethological imperatives which obscured any socio-political interpretation. For a recent attempt to recover and recast the debates on human territoriality, see Sack, 1986. Neither my earlier work nor Sack's, however, provide a satisfactory social ontology of territoriality.

turn the setting for a contingent regionalism, an active consciousness and assertiveness of particular regions, *vis-à-vis* other regions, as territorial and social enclosures. As an expression of the territoriality of locales, regionalism is grounded in the geography of power.

Material being in the form of the body is the first and prefigurative instantiation of this hierarchy of differentiated nodal locales. Ego is the primal tension-filled centering of being, and around it is formed a created regionalization that has escaped formal analysis until very recently, for it remained stubbornly outside what Giddens describes as our discursive (as opposed to practical) consciousness. Giddens turns primarily to Goffman's sociology of encounters and Hagerstrand's time-geography for insight into this ego-centred regionalization, but equally insightful conceptual salvoes can be found in the work of Edward Hall, Robert Sommer and others who helped to spatialize the ego through a cultural critique and the initiation of an environmental psychology of spatial cognition (Soja, 1971). What has become increasingly revealed in these writings is a remarkable micro-geography of human interaction hingeing around the portable bubbles of personal space zonation and 'proxemic' behaviour, a non-verbal and unwritten ordinary language of spatial intersubjectivity.

But this is only the beginning, the first of many layerings of created locales and regionalizations rippling outward from the subjective spatiality of the portable ego to become imprinted on the humanized landscape. Nodality twines together collective activities around other centred and relatively fixed settings which are also regionalized and more or less territorially bounded. In the modern world, the place of work and the place of residence are the predominant nodal locales of social co-presence and their locational separation and territoriality induces its own distance-ordered but socially produced patterning of human interaction and experience. In less modern contexts, these two locales are typically co-centred and reinforce one another to define more tightly bounded enclosures of social integration relatively impermeable to interaction at higher geographical scales, except through the agglomeration of nodal locales and individual micro-geographies into human settlements or what might usefully be called *localities*.

Localities – another term which Giddens does not use – can be defined as particular types of enduring locales stabilized socially and spatially through the clustered settlement of primary activity sites and the establishment of propinquitous territorial community. Like every locale, they are spatio-temporal structurations arising from the combination of human agency and the conditioning impact of pre-existing spatio-temporal conditions. They provide another created setting, a more elaborate built environment, for human interaction expanded in scale, density, social differentiation, and collective attachment to place. They are also generative locales for what Giddens defines as 'distanciation', the stretching of social systems over time–space from the co-presence of local social integration to the more encompassing and elastic collectivities and reciprocities of system integration. Localities are thus the building blocks of urbanization: the formation of nodally clustered and cohesive locales regionally differentiated internally (within the cluster), comparatively (one urbanized locale versus another), and hierarchically (positioned within a multi-level system of urban locales). Towns and cities may be described as localities which encompass contexts, enclosures and nodal concentrations of human interaction which are linked to both social and system integration, and hence to multiple networks of social power. In the context of the contemporary world, the locality can range from the smallest settlement or neighbourhood to the largest conurbation.

Urbanization, however, represents a break from ontological generality and forces a transition to a more concretely specified historical geography, a shift which Giddens fails to make sufficiently explicit. Every human society that has existed has been contextualized and regionalized around a multi-layered nesting of supra-individual nodal locales – a home-base for collective nourishment and biological reproduction, collection sites and territories for food and materials, ceremonial centres and places to play, shared spaces and forbidden terrains, definable neighbourhoods and territorial enclosures. But only in some societies have these locales been agglomerated into specifically urban settlements and only in the past two centuries has urbanization expanded to become the dominant life setting for a major portion of the world's population, even in areas conventionally defined as 'non-urban' or rural. This is the extended definition of the urban used by Lefebvre to describe the specific geography of capitalism.

Understanding urbanization and urbanism in the contextuality of hierarchically centred locales thus projects rather than rejects the Giddensian ontology. Giddens does not succeed in developing a rich and rigorous theory of urbanization, choosing instead to focus his projections on the nation-state (as if the state supplants rather than embodies urbanization as the primary locus of power). But he does insist on locating the urban at the heart of critical social theory and in the midst of time–space structuration. The specificity of the urban, that old question which so divided Marxist geographers and sociologists, is thus given a new look and significance.

Urbanization can be seen as one of several major accelerations of time–space distanciation that have extended the scale of human interactions without necessarily destroying their fundamental spatial anatomy. You and I still live within a hierarchy of nodal regionalizations emanating from our bodies, but social interaction and societal integration have now expanded to a world scale, a global reach in which the urbanization process has been a primary vehicle. The specificity of the urban is thus defined not as a separable reality, with its own social and spatial rules of formation and transformation; or merely as a reflection and imposition of the social order. The urban is an integral part and particularization of the most fundamental contextual generalization about the spatiality of social life, that we create and occupy a multi-layered spatial matrix of nodal locales. In its particularity, its social specificity, the urban is permeated with relations of power, relations of dominance and subordination, that channel regional differentiation and regionalism, territoriality and uneven development, routines and revolutions, at many different scales.

The descriptive generalities of the Chicago School and most of modern urban sociology and geography – claiming that cities are distinguished (presumably from the rural or non-urban) by their size, density, heterogeneity, anomie, functional solidarities, geographical concentricities and axialities – are not inaccurate. But they conceal the more fundamental specificity of the urban that arises from the conjunction of nodality, space, and power. Cities are specialized nodal agglomerations built around the instrumental 'presence availability' of social power. They are control centres, citadels designed to protect and dominate through what Foucault called 'the little tactics of the habitat', through a subtle geography of enclosure, confinement, surveillance, partitioning, social discipline and spatial differentiation.

The ability to control emanates in large part from nodality/centrality itself and extends outwards along at least two planes, one directly from centre to hinterland (a vicinal control which typifies social integration) and the second from one nodal centre to others (a hierarchical control characteristic of system integration). Together these urban and territorial emanations of power and control define the very nature of the state. They also define a contestable terrain of spatial politics and civic struggle over *le droit à la ville*, the 'rights to the city' in Lefebvre's terms, the power of citizens to control the social production of space.[10]

As Giddens writes, the city is 'far more than a mere physical *milieu*. It is a "storage container" of administrative resources' around which states

---

10. See Michael Mann, *The Sources of Social Power* (volume 1, 1986) for the beginnings of what promises to be one of the few explicitly geographical analyses of the state and social stratification. Mann starts out with the following underlined assertion: '*Societies are constituted of multiple overlapping and intersecting sociospatial networks of power*' (1). He goes on to note that 'Most theorists prefer abstract notions of social structure, so they ignore geographical and sociospatial aspects of societies. If we keep in mind that "societies" are *networks*, with definite spatial contours, we can remedy this' (9).

are built (1984, 183). He notes the dramatic shifts in the contextuality of the city which come about with the rise of capitalist industrialization and its commodification of time and space at the end of a chapter on 'Time, Space and Regionalization'. He then turns, appropriately enough, to Foucault for critical insight into the 'timing and spacing' of disciplinary power, building these insights into a subsequent analysis of the structural principles of Tribal, Class-Divided, and Class (capitalist) Society. Here an important distinction arises between the dominant locale organization of class-divided societies, rooted in the symbiosis of city and country-side, the 'axis relating urban areas to their rural hinterlands'; and the dominant locale organization of capitalism, the 'sprawling expansion of a manufactured or "created environment"'. How close this is to focusing our attention on the problematic and instrumental spatialization that has marked the historical geography of capitalism, that Lefebvre unmasked and tied so closely with urbanization, that others have begun to identify as the key to understanding contemporary capitalist society.

But Giddens again spirals right up to the edge of the Lefebvrean version, only to refuse to take the next (lateral?) step. In the second half of *Constitution*, the Giddensian helix begins an almost regressive involution, repeating itself without advancing very far forward. The vivid and central significance of spatiality seems to be stripped away piece by piece until we are left, in a long chapter on the application of structuration theory to empirical research and social critique (which follows another exorcism of Talcott Parsons!) with almost no space at all. There is a brief mention of uneven development as having 'a broader application than has ordinarily been recognized' (319), followed by several, almost Wallersteinian, sentences on regionalization both producing and diffusing social contradictions. But the explicit advice given to the 'social analyst' seems to have omitted the forceful assertion that 'space matters' after all. It is no wonder that the sociological response to Giddens, both pro and con, has almost entirely failed to recognize the significance of his pronounced spatial turn, for Giddens himself seems to cover it up at the most critical moments.

Giddens, at the very end of *Constitution*, tries hard to recover his geography after stashing it away for the previous 150 pages. Posed as the ultimate afterthought, he writes:

> The phrase might seem bizarre, but human beings do 'make their own geo-graphy' as much as they 'make their own history'. That is to say, spatial con-figurations of social life are just as much a matter of basic importance to social theory as are the dimensions of temporality. (1984, 363)

These are the last sentences of the text:

Space is not an empty dimension along which social groupings become structured, but has to be considered in terms of its involvement in the constitution of systems of interaction. The same point made in relation to history applies to (human) geography: there are no logical or methodological differences between human geography and sociology! (368)

Whether these terminal and exclamation marked statements on 'Social Science, History and Geography' will remain Giddens's last words on the subject or become the seeds for another spiralling sequel, it is difficult to foretell.

Looking back over *The Constitution of Society*, there remains much to be praised. In my view, the infusion of power into an explicitly spatialized ontology of society and hence into interpretations of the making of geography alongside the making of history is Giddens's major achievement. Similar arguments are there in the work of Foucault, Lefebvre, Poulantzas, Sartre and perhaps others I have missed. But in *Constitution*, Giddens brings almost everything together in a monumental synthesis which provides, for the first time, a systematic social ontology capable of sustaining the reassertion of space in critical social theory.

The easiest criticism of Giddens is the most complicated and possibly the most futile, for he has set up such strong armour against it in his personalized evolution. By leaving epistemology to others and concentrating on social ontology, Giddens frees himself to dip into empirical analysis at will and without commitment to any but his own framework of interpretation, his own 'cosmic determination' (to capture again the head quotation from Sartre). This is, of course, not uncommon among the best social theorists and philosophers. But it makes Giddens vulnerable to missing the particularities of the contemporary moment, its new possibilities and its breaks with the past. As the sociologist and theoretical realist John Urry writes:

[Giddens] tends to neglect the problems of explaining the causes and consequences of recent transformations in the spatial structuring of late capitalism. Moreover ... this omission is particularly serious since it is space rather than time which is the distinctively significant dimension of contemporary capitalism, both in terms of its most salient processes and in terms of a more general social consciousness. As the historian of the *longue durée*, Braudel, argues 'All the social sciences must make room for an increasingly geographical conception of mankind'. (1985, 21)

This is essentially a call for both a more empirical and more spatially centred application of critical social theory to the perplexities of the present day. And it takes us into another round of *restructuring*, a deeper and more radical deconstruction and reconstitution of critical social theory than Giddens has apparently contemplated.

Making sense of contemporary modernity, or postmodernity if you will, cannot be done by simply announcing the logical and methodological equivalence of history, geography, and sociology in their Modernist guises, and extolling the fruitfulness of their nascent reconnections. The whole fabric of the Modern academic and intellectual division of labour that has defined, enclosed, and reified these disciplines since the late nineteenth century must be radically reshaped. Giddens's residual sociologism thus takes on a new significance, for the Sociology that has been so richly consolidated and expanded by Giddens stands today as one of the many reified disciplinary monuments that need to be deconstructed before we can successfully do anything new.

# 7

# The Historical Geography of Urban and Regional Restructuring

> Capitalist development must negotiate a knife-edge between preserving the values of past commitments made at a particular place and time, or devaluing them to open up fresh room for accumulation. Capitalism perpetually strives, therefore, to create a social and physical landscape in its own image and requisite to its own needs at a particular point in time, only just as certainly to undermine, disrupt and even destroy that landscape at a later point in time. The inner contradictions of capitalism are expressed through the restless formation and re-formation of geographical landscapes. This is the tune to which the historical geography of capitalism must dance without cease. (Harvey 1985, 150)

David Harvey's restatement of his earlier depiction of urbanization and struggles over the built environment signalled a shift in the scope of his arguments, from the specifically urban to a more comprehensive geographical and analytical landscape. Projected into this larger landscape is a dynamic and contradiction-filled dialectic of space and time, human agency and structural constraint, a historical geography that is played out at many different scales, from the routinized practices of everyday life to the more distant geopolitical shufflings of a global spatial division of labour.

The landscape has a textuality that we are just beginning to understand, for we have only recently been able to see it whole and to 'read' it with respect to its broader movements and inscribed events and meanings. Harvey's inaugural reading focuses on the hard logics of the landscape, its knife-edge paths, its points of perpetual struggle, its devastating architectonics, its insistent wholeness. Here, capital is the crude and restless *auteur*. It strives and negotiates, creates and destroys, never fully able to make up its mind. Capital is seen as two-facedly choreographing the chronic interplay of time and space, history and geography, first trying to annihilate with temporal efficiency the intransigent social physics of space only to turn around again to buy time to survive from the very spatiality that it seeks to transcend. This stressful ambivalence etches itself everywhere, organizing the landscape's material forms and configurations in an oxymoronic dance of destructive creativity. Nothing is wholly determined, but the plot is established, the main characters clearly defined, and the tone of the narrative unshakably asserted.

The real text, of course, is much more subtly composed and filled with many different historical and geographical subtexts to be identified and interpreted. Capital, above all, is never alone in shaping the historical geography of the landscape and is certainly not the only author or authority. Yet the landscape being described must be seen as a persistently capitalist landscape with its own distinctive historical geography, its own particularized space-time structuration. The initial mapping, at least, must therefore never lose sight of the hard contours of capitalism's 'inner contradictions' and 'laws of motion' no matter how blurry or softened history and human agency have made them. The plot has thickened but not enough to obliterate an enduring central theme.

The 'restless formation and reformation of geographical landscapes' that is triggered by the dynamics of capitalist development has been the most important discovery arising from the encounter between Western Marxism and Modern Geography. In the preceding chapters, I have tried in many different ways to communicate the significance of this problematic spatiality and to begin its practical and theoretical exploration – that is, to outline the foundations for a politically-charged empirical interpretation of the historical geography of capitalism. For these purposes, the term 'spatialization' has been applied in both a general sense, to indicate the increasing reassertion of a spatial emphasis in ontological, epistemological, and theoretical discourse and in our practical understanding of the material world; and as a particular phenomenological and ideological process associated with the development and survival of capitalism as a mode of production, a political economy, and a material culture of modernity. The general and particular meanings of spatialization revealingly intersect in the contemporary context, in the formation of postmodern geographies, but do not lose their separable qualities.

It is now possible, however, to begin to generalize about the particularities of capitalist spatialization and to particularize the generalizations of a spatialized critical social theory. The two projects intertwine today primarily in the analysis of urban and regional restructuring, or to use the terminology developed in the previous chapter, the changing configuration and political meaning of the nested hierarchy of regionalized and nodal locales that has been evolving in distinctive ways since the origins of capitalism. This encompassing and institutionalized spatial matrix is currently in the midst of a dramatic reformation that is itself both generalizable and unique, evocative of past periods of prolonged crisis and restructuring, yet filled with new conditions and possibilities that challenge established modes of understanding. Making theoretical and practical sense of this contemporary restructuring of capitalist spatiality has become the overriding goal of an emerging postmodern critical human geography.

## Observations on the Concept of Restructuring

Restructuring, in its broadest sense, conveys the notion of a 'brake', if not a break, in secular trends, and a shift towards a significantly different order and configuration of social, economic, and political life. It thus evokes a sequential combination of falling apart and building up again, deconstruction and attempted reconstitution, arising from certain incapacities or perturbations in established systems of thought and action. The old order is sufficiently strained to preclude conventional patchwork adaptation and to demand significant structural change instead. Extending Giddens's terminology, one can describe this brake-and-shift as a time-space *re*structuration of social practices from the mundane to the *mondiale*.

These societal restructuring processes continue to be buried under idealized evolutionary schemata in which change just seems to happen, or arises to punctuate some ineluctable march towards 'progress'. This evolutionary idealism (another form of historicism) disguises the rootedness of restructuring in crisis and in the competitive conflict between the old and the new, between an inherited and a projected order. Restructuring is not a mechanical or automatic process, nor are its potential results and possibilities pre-determined. In its hierarchy of manifestations, restructuring must be seen as originating in and responding to severe shocks in pre-existing social conditions and practices; and as triggering an intensification of competitive struggles to control the forces which shape material life. It thus implies flux and transition, offensive and defensive postures, a complex and irresolute mix of continuity and change. As such, restructuring falls between piecemeal reform and revolutionary transformation, between business-as-usual and something completely different.

That we are currently involved in an ongoing period of intensive societal restructuring seems, with the increasing clarity of hindsight, difficult to deny. There is also widespread agreement among those attempting to interpret this contemporary restructuring that it was sparked by a series of interrelated crises – from the urban insurrections of the 1960s to the deep worldwide recession of 1973–75 – which marked the end of the prolonged period of capitalist economic expansion following the Second World War. Although less widely agreed upon, it can also be argued that these crises arose primarily in conjunction with the particular institutional structures which sustained and shaped the expansionary capitalist accumulation of the post-war boom years. More specifically, they can be seen as a complex chain of crises: in the established international division of labour and global distribution of political and economic power; in the expanded and now clearly contradictory functions of the national state; in the Keynesian welfare systems and stabilizing social contracts between governments, corporations, and organized labour; in the patterns of uneven regional development that had become so firmly established within countries over the preceding century; in the developed forms of exploitation of women, minorities, and the natural environment; in the spatial morphology, industrialization, and financial functioning of cities and metropolitan areas; in the design and infrastructure of the built environment and collective consumption; in the ways in which capitalist production relations are imprinted into everyday life, from the labour process in the workplace to the reproduction of life, labour, and patriarchal power in the household and home.

Each of these arenas of crisis and restructuring has spawned a growing academic literature and new contexts for critical debate. Each has also generated its own constellation of popular catchphrases trying to capture and define the new forms assumed to be emerging. One of the earliest and most familiar announced the birth of a 'Post-Industrial Society', with its telescoped end to the industrial era and its associated ideologies in the advanced capitalist countries, and the concurrent shift to a new service-based economy as an idealized conclusion to the epochal climb through progressive 'stages of growth'. However, there are many others which by now have filled the popular imagination with projected epithets of restructuring: a 'New International Economic Order' induced by the 'industrial miracles' of the NICS (the 'Newly Industrialized Countries') and the 'Peripheralization of the Core' in an 'Age of Global Capital' recentred around the 'Pacific Rim'; a 'Power Shift' pitting 'Sunbelt versus Frostbelt' arising from the 'Deindustrialization of America' and the growth of 'High Technology' and the 'Electronics Revolution' in new 'Silicon Landscapes'; 'Post-Welfare States' struggling with the 'New Austerity' to create an 'Information-based Society' and 'Post Fordist' industrial systems.

Picking a way through this metaphorical forest is no easy task. Each of these labels helps somewhat to illuminate the restructuring process, but they often seem to flash so brightly that they prevent us from seeing what may actually be taking place in all its fulsome complexity and inter-contingency. It becomes necessary, therefore, to present a brief reading of where I intend to go in the analysis and interpretation of contemporary restructuring processes.

The starting-point is the assertive connection between restructuring and spatialization. The contemporary moment will thus be looked at as the most recent attempt to restructure the spatial and temporal matrices of capitalism, another search for a survival-aimed spatio-temporal 'fix'. If there is to be a historico-geographical materialism (or, if one prefers, a robust critical human geography), it will come from making theoretical and practical sense of this contemporary spatial, temporal, and societal restructuring.

To emphasize the most conventionally neglected element in this threefold restructuring (and to defend against its submergence in the other folds and creases), I will focus on three major streams of spatial restructuring. The first begins with the ontological restructuring that has advanced the reassertion of space in critical social theory and philo-sophical discourse, the main theme of the previous chapters. If we need a contemporary epithet for this theoretical restructuring, let it be 'post-historicism'. The second stream has been carried along by the spatializ-ation of Western Marxism and concerns the material political economy of capitalist accumulation and class struggle in the context of urban and regional development. For the present moment, 'postfordism' provides a convenient capsulization of this stream of spatial restructuring. The third stream adds to urban and regional political economy an insistent cultural dimension and critique which stretches restructuring into debates on the nature of modernity, modernization, and modernism. Here the argu-ments are caught in the assertion of a 'postmodernism' rising Janus-like in response to the pervasive instrumentalities of the contemporary moment.[1]

The three streams will come together most propitiously in the empirical context of an exemplary place, the confluent city-region of Los Angeles. Before looking at Los Angeles, however, I will venture primarily along the second stream to describe some aspects of the politi-cal economy of urban and regional restructuring, its spatiality, period-

---

1. As used here, the asserted prefix *post* indicates 'following upon' or 'after' but does not mean complete replacement of the modified term. Modernism, Fordism, and historic-ism persist even after their post–defined restructuring. Similarly, certain features of post-modernism, postfordism, and posthistoricism can be found prefiguratively in a more distant past. This usage may confound those who insist upon the application of a categorical logic of temporality and sequence, but that is partly its intent.

icity, and historical geography. It will be a rather sweeping journey without the thick empirical detail and complexity that are necessary to give depth and conviction to the arguments, especially perhaps to the idiographically-minded historian and geographer. But it indicatively sets the scene as a suggestive outline aimed at stimulating further historico-geographical inquiry.

## Regions in Context: on 'Restructuring and the 'Regional Question'

During the past twenty years, significant changes have been occurring in the patterns of uneven regional development that had become so firmly established within advanced capitalist countries over the preceding century. In much the same way that concurrent developments within the global economy seem to be shattering the neat compartmentalization of First, Second, and Third Worlds and inducing proclamations of a 'New International Division of Labour', so too has the patterned mosaic of subnational regional differentiation become more kaleidoscopic, loosened from its former rigidities. The accelerating shift in industrial production and political power from the Frostbelt to the Sunbelt in the USA, for example, has become a sweeping metaphor enclosing and structuring public debate and scholarly research on the 'regional question'. Similarly, attention has been drawn to an emergent 'Third Italy' (Bagnasco, 1977) complicating the simple dualism that had heretofore defined the most classical 'North–South' model of uneven regional development; and to a series of 'role reversals of regions' in many other countries, as once prosperous industrial areas decline in tandem with the rapid industrialization of formerly less developed regional peripheries.

As many have been arguing, the regional question and the analysis of regional restructuring have been placed on the contemporary political and theoretical agenda with renewed force. Consider, for example, the titles of some recent publications: 'Regions in Crisis', 'Capital Versus the Regions', 'The Regional Problem', 'Regions in Question', 'Regional Wars for Jobs and Dollars', 'Regional Analysis and the New International Division of Labour', 'Regionalism and the Capitalist State', 'Uneven Development and Regionalism', 'Global Capitalism and Regional Decline', 'In What Sense a Regional Problem?', 'The North Will Rise Again', 'Regional Development and the Local Community', 'Profit Cycles, Oligopoly, and Regional Development'. Clearly something is astir, as alternative views jockey for position in an emerging regional political economy, each seeking an appropriate understanding of contemporary regional change.

Four interpretive contexts can be used to situate the contemporary debates on the regional question. The first, and most comprehensive, is the transformative retheorization of space, time and social being, that is currently taking shape in contemporary social theory and philosophy. Little more need be said about this situating context other than to emphasize again that it springs primarily from a reconstructed ontology of human society in which the formation of regions, the patterning of uneven regional development and regionalism, and the formulation of regional theory are directly implanted in an encompassing process of spatialization, the social production of space. Concrete as well as concretizing, historically situated and politically charged, this spatial structuration of society gives an interpretive specificity to regions as part of a multilayered spatiality which stretches in its impact from everyday life in an immediate built environment of social integration to the systemic networks of flows and transactional connections that tie together the global space economy. Subnational regions are thus amongst the many created and constitutive locales of social life, contingent upon social and historical processes while simultaneously formative of society and history.

The second interpretive context arises from a more specific retheorization of the causes and consequences, the nature and necessity, of geographically uneven development. As previously argued (Chapter 4), geographically uneven development is an essential part of capitalist spatiality, of its distinctive spatial matrix and topology. Produced and reproduced at multiple scales, it is inherent to the concretization of capitalist social relations and regionalizations both as medium/presupposition and as outcome/embodiment. Like spatiality itself, geographically uneven development has traditionally been seen as an external reflection of social forces, an illusive mirror of social action and the struggle of social classes. It, too, is now being appropriately reconceptualized in a reconstructed historical and geographical materialism.

Situating the (subnational) regional question in the context of geographically uneven development connects it with the dynamics of changing spatial divisions of labour and with the interplay of regionalization and regionalism. Subnational regions defined in this way are thus the product of a regionalization at the level of the national state, a particularized geographical differentiation that is as tentative, ambivalent, and creatively destructive as any other component of the spatial matrix of capitalist development. Similarly, this subnational spatial division of labour may provide effective channels of exploitation, or it may not – it has no automatic and predetermined functionality to the logic of capital. It is a resultant spatialization arising from competitive struggles and particular conjunctures, filled with tensions, politics, ideology, and power. Regionalism in turn is a possible response to regionalization, a 'reaction formation' to borrow a term used to describe ethnicity and other communal identities. Regionalism can take on many different political and ideological forms, ranging from an acquiescent request for additional resources to an explosive attempt at secession (Hadjimichalis, 1986).

These dynamics of subnational regionalization and regionalism cannot be easily generalized – or specified – for they are so essentially conjunctural and are periodically substantially restructured. Hence the need to concretize the regional question still further by placing it within a third interpretive context: the periodicity of regionalization in the historical geography of capitalism. This takes us back into Harvey's depiction of the restless formation and reformation of geographical landscapes and Mandel's 'long wave' periodization of capitalist development. As noted earlier, Mandel's interpretation of the regional question springs from his assertion that the whole capitalist system appears as a hierarchical structure of different levels of productivity – as 'the uneven development of states, regions, branches of industry and firms, unleashed by the quest for superprofits' (1975, 102). This search for superprofits centres around three major sources, two defined primarily around regional differentiation (subnational and international) and the third around sectorally uneven development. All three sources have existed since the origins of capitalism, but Mandel argues that each has achieved particular prominence in different historical periods.

During what he calls the age of 'freely competitive capitalism' (up to the end of the nineteenth century) the dominant form of superprofits derived from the regional juxtaposition of industry and agriculture within the then advanced capitalist countries, a juxtaposition that was deeply imbricated in the relations between city and countryside. Industrial capital and production were concentrated and localized in only a few territorial complexes, surrounded by rings of agrarian regions serving to supply raw materials and food, markets for industrial consumer goods, and reservoirs of cheap labour.

This distinctive regional division of labour was consolidated through the formation of integrated national markets (as in the unification of Germany and Italy in the late nineteenth century) and reinforced hegemony over dependent territories. The classic case of an agrarian 'subsidiary country' was Ireland, whose budding industries, as Marx noted, were destroyed in an early example of the process of capitalist underdevelopment. Other similar subsidiary regions or 'internal colonies' included Flanders, the American South, the Italian Mezzogiorno, many parts of the Austro-Hungarian Empire, some eastern and southern sections of Germany (Bavaria, Silesia, Pomerania-Mecklenburg), the agrarian west and centre of France, and Andalucia. What was occurring throughout the nineteenth century was thus a regional restructuring and expansion in scale of the preformative town–countryside relationship and the 'primitive accumulation' which marked the origins of capitalism. The geographical expansion of commercial capital initially paved the way, but the dominant spatializing force was associated with massive urban industrialization.

Nineteenth–century regionalisms developed mainly out of attempts to preserve distinct regional cultures in the face of increasing homogenization or to resist the particular spatial divisions of labour being imposed by expansive market integration and the equally expansive national state. For the most part, this involved subsidiary agrarian regions, but relatively industrialized locales such as Catalonia also responded. Anarchist thought, with its explicit antistate and decentralist principles, found fertile ground in many of these regional territorialities and provided the major radical threat to urban–industrial capitalism during most of the century. But within the national peripheries there also developed powerful new regional hegemonic blocs (to borrow from Gramsci) which welcomed, orchestrated, and benefited from the processes of regional underdevelopment. They helped to consolidate the regional structures of power and to subdue more radical regionalisms.

In the age of imperialism and the rise of corporate monopolies and oligopolies, the primary source of superprofits began to shift. As part of another scale-expanding and crisis-induced restructuring hingeing around the *fin de siècle*, the international juxtaposition of development (in the imperialist states) and underdevelopment (in colonial and semicolonial territories) became more important to the survival of capitalism than subnational regional differentiation. Superexploitation of a newly consolidated global periphery spurred recovery from the late–nineteenth–century depression and led the rapid expansion that occured in the core countries during the first two decades of the twentieth century. Capitalism did not suddenly internationalize. Mercantile capital had been operating to extract superprofits throughout the world for centuries through commodity trade. Imperialism, however, internationalized another circuit of capital, involved in finance, money, and investment transactions, which more efficiently organized the international economy for larger scale geographical transfers of value than had ever before been possible. The old city–countryside relationship became implanted not only on a national scale but also in a global structure of capitalist core and periphery.

Regional underdevelopment in the core countries did not disappear, nor did the pressure of antagonistic regionalisms. But there was some significant regional restructuring, shaped largely by the geographically uneven impact of internationalization and imperialist profit-takings, and by the accelerated concentration and centralization of domestic capital (exemplified by the rash of vertical and horizontal mergers around the turn of the century, especially in the USA). Regions containing the main imperialist capitals and the centres of monopoly control (for example, major corporate headquarters) tended to grow much more rapidly than those which may have formerly been at similar levels of industrial development but did not benefit as much from global profit-taking. The major scale for spatial restructuring in the core countries, however, was not international as much as it was intra-urban. The 'inherited' regional divisions of labour within countries probably did not change very much during this period, even when the overall intensity of regional inequalities was significantly reduced. Indeed, this relative stability, if not amelioration, of the structure of uneven regional development was itself an indication of the shrinking relative importance of the superprofits derived from subnational regional differentiation.

The older agrarian peripheries were either partially urbanized or left relatively alone, but their key role in supplying cheap labour, food, raw materials, and markets was increasingly transposed to the 'external' colonies. Where these functions did remain important, such as in Ireland and the Mezzogiorno, peasant rebellions and aggressive regional movements did arise (as did massive outmigration). But the main political and economic action was hierarchically sandwiched around the regions, in the major urban centres 'below' and in the world-scale arena of inter-imperialist rivalry 'above'.

This distinctive regime of accumulation, based on the consolidation of an international division of labour partitioned into a dominant-industrial-imperialist world core and a dependent-agrarian-underdeveloped world periphery, as well as on the more centralized corporate structure of monopoly and finance capital, traced a similar path of historical development to that of the competitive–entrepreneurial regime that preceded it. Emerging clearly in the period of crisis and restructuring of the late nineteenth century, it became the foundation for the expansive boom in the early twentieth century, only to plunge into its own crisis and restructuring phase during the Great Depression. Monopoly capitalism did not disappear, however, just as its assertion as a predominant regime of accumulation did not erase its predecessor. What developed was another 'layer', a reorganized regime of accumulation articulated with its residual antecedents and able to co-ordinate recovery from the deepest depression in capitalist history/ geography.

Mandel describes this new regime of accumulation as 'Late Capitalism' and argues that its appearance marks a shift in the primary source of superprofits from geographically uneven development to 'the overall juxtaposition of development in growth sectors and underdevelopment in others, primarily in the imperialist countries but also in the semi-colonies in a secondary way' (Mandel, 1975, 103). As he is careful to note, these technological rents – profits originating from advances in productivity based largely on technological developments and the organization of production systems – existed in earlier periods and were essential to the very origins of capitalism. In the absence of high levels of centralization and concentration of capital, however, the appropriation of technological rents tended to be limited in magnitude and of short duration, especially given uncontrolled entrepreneurial competition. Only within Late Capitalism, he argues, do they become predominant and efficiently systemic. The rising importance of technological rents overshadows somewhat the preeminent role played by the exploitation of geographically uneven development in previous regimes of accumulation. But at the same time, they become more central to an understanding of the changing regional divisions of labour and the changing nature of the regional question over the past fifty years. As the sectoral sources of superprofits rise, the spatial sources do not necessarily decline to insignificance, for geographically uneven development can be continuously reconstituted. Indeed, the contemporary search for superprofits, wherever they can be found, is reaching a competitive pitch that is higher than ever before in the history of capitalist development, recreating new and more complex patterns of development and underdevelopment all over the world.

Muted during the Great Depression and in the immediate post-war period, the regional question took on new importance in the 1950s. From the reorientation of the British New Towns programme towards more regional versus exclusively urban problems, to the formation of DATAR in France and the beginnings of spatial planning for the Italian Mezzogiorno and other 'backward' regions in core countries, regional welfare planning was put on the public agenda, often but not always in response to regional unrest and regional political movements. As was true for other forms of planning, and the role of the state more generally, there were two potentially contradictory sides to these promises of improved regional welfare under the aegis of what can be called a regime of state-managed capitalism. Centralized state planning required a sustained social legitimization, especially from those most likely to create political and economic disorder. Promising more balanced regional development accorded well with this objective. At the same time, the state – in no small part because it depended upon tax revenues – had also to facilitate the capitalist accumulation process, which did not consistently coincide with regional welfare improvements. After all, uneven regional development has always been an important foundation for the generation and extraction of superprofits and continues to be so even in an age when the primary source of superprofitability may have shifted to sectorally uneven development.

This contradictory role of the welfare state and of regional welfare planning remained relatively invisible well into the 1960s, although virtually every major regional development programme aimed at reducing regional economic inequalities met with powerful resistance from some segments of private capital. Even when their initiation could not be blocked, programme activities were at least partially co-opted to benefit highly centralized and concentrated capital interests, often draining resources (and available technological rents) from the 'backward' region into the most developed areas of the national space economy. This was true for the main regional welfare planning experiment in the USA, centred on Appalachia, and was repeated over and over again in France (for Brittany and the south), in Britain (for nearly all the designated regional development areas), in Italy (via the Casa per il Mezzogiorno). A similar pattern also operated globally in the 'backward' peripheral countries which increasingly adopted the systematic spatial planning models originating in the core and often promoted by core countries as a postcolonial planning panacea.

Awareness of the countervailing spatial strategies of capital (and of the state) was minimal on the part of both the targeted regional poor and the experts (theoreticians and practitioners) who were shaping regional policy. For the most part, the first group were soothed by the promises proffered even when they were not immediately delivered. At least someone seemed finally to be aware of the regional problem and was proposing to do something about it. The regional planners, enthralled by the new spatial theories of the time, were also inclined to be blissfully accommodative, convinced that their idealistic objectives could be obtained through good intentions and innovative spatial planning ideas.

Regional planning never received especially large public expenditures, but the period from 1950 to 1970 was a golden age of sorts in the history of regional development theory and practice (Weaver, 1984). 'Growth poles' and 'growth centres', regional science and spatial systems analysis, urban systems modelling and other efforts aimed at reconstructing a more balanced and equitable hierarchy of nodal regions, were at the height of popularity. By 1970, virtually every country in the world had adopted some form of spatial planning policy, in some instances placing it as the centrepiece of the national economic development plan. This worldwide expansion of a Keynesian regional planning signalled an explicit, if often only documentary, commitment by the state to redress regional (and international) inequalities, in effect to change the established spatial division of labour. As long as the promises appeared potentially achievable, antagonistic anti-state regionalisms remained relatively quiescent, waiting for the goods to be delivered.

The series of crises and recessions which marked the end of the postwar economic expansion of core capitalism also destroyed the patient optimism of regional welfare planning. New sectoral and spatial 'liberation' movements arose to challenge the orderly processes of accumulation. Radical nationalist movements (including the Chinese Cultural Revolution) removed more than a billion people from the profitable games of international development and underdevelopment or at least significantly disrupted the smoothness of the play. A resurgence of regionalisms similarly strained the established operations of the welfare states, while new places began to compete successfully to localize, if not liberate, the technological rents being generated by the most propulsive sectors of an increasingly flexible, postfordist, and seemingly 'disorganized' regime of capitalist accumulation (Piore and Sabel, 1984; Scott and Storper, 1986; Offe, 1985; Lash and Urry, 1987). The sequence of spatializations that has marked the historical geography of capitalism moved into another round of crisis and restructuring, churning up a new set of competitive possibilities for reformation and transformation. To continue to make sense of the regional question thus requires a fourth context of interpretation, one that is grounded in the particularities of the contemporary restructuring process.

Not all regional political economists today depend upon the specifically Mandelian conceptualization of long waves, Late Capitalism, and technological rents in their analyses of the contemporary period. Yet there is a remarkably similar emphasis and interpretation evident within the wide range of contemporary regional approaches, from the study of the articulations of 'submodes' of production and the rise of 'global capitalism' (Forbes and Thrift, 1984; Gibson and Horvath, 1983; Gibson et al., 1984); to the French modes of regulation/ regimes of accumulation school and its arguments about the 'globalization of the crisis of Fordism' (Aglietta, 1979; Lipietz, 1984a, 1984b, 1986); to the new industrial geography of territorial labour markets, sectoral profit cycles, high-technology-based industrial complexes, and changing spatial divisions of labour (Massey, 1984; Walker and Storper, 1984; Cooke, 1983; Scott and Storper, 1986). There are differences in periodization, terminology, and especially in the specific emphases given to the political implications of contemporary restructuring, but there are significant commonalities that reinforce rather than reduce the strengths of an emerging historico-geographical materialist perspective on the regional question.

All share a similar crisis model of historical and geographical change; an emphasis on class analysis and the labour process; an appreciation for the relevance of technology and corporate structure in the differentiation of productivity and profits; explicit attention to the interplay between spatiality, politics, and the role of the state; a concern for analysing the internationalization of capital and the associated acceleration of capital mobility and labour migration; and a vision which recognizes, to varying degrees, the general nature and particular distinctiveness of capitalist spatialization. More specific interpretations typically hinge around a historical turning point in the late 1960s or early 1970s and its echoing of Great Depressions of the past; and there is an open acceptance of the general restructuring hypothesis: that we are currently in the midst of a period in which capital and labour are being significantly reorganized in an attempt (not yet completely successful) to restore rising profits and reinforce labour discipline, in part through direct attacks on working-class organization, wages, and standards of living.

There is also general agreement that the 'new' regional and international divisions of labour taking shape over the past twenty years are not total replacements of the 'old' divisions, which remain not only alive, but kicking as well. The historical geography of capitalism has not been marked by grand turnabouts and complete system replacements, but rather by an evolving sequence of partial and selective restructurings which do not erase the past or destroy the deep structural conditions of capitalist social and spatial relations. There is thus no justification for a naive and simplistic 'rush to the post' – to a postindustrialism, post-capitalism, postMarxism – that insists on a finalizing end to an era, as if the past can be peeled away and discarded.

Nevertheless, there have been significant regional changes taking place in the current period of restructuring and these need to be noted. Paradoxical as it may initially appear, one noteworthy change has involved an intensification of pre-existing patterns of uneven regional development in many areas, and a consequent reinforcement of old core and periphery divisions. Several well-established core regions have experienced sustained and even expanded relative economic and political power, while many backward peripheries have plunged deeper into relative impoverishment, in some cases into pandemic famine. These 'intensified continuities', however, are not simply more of the same, for they have been occurring under a new set of sectoral, social, political, and technological conditions that are significantly modifying how geographically uneven development is produced and reproduced. Identifying and understanding these altered conditions has become the critical focus for contemporary interpretations of regional restructuring.

To illustrate, it is useful to return to the technological rents and sectorally uneven development which Mandel has argued have supplied the major source of superprofits since the Second World War. As noted earlier, technological rents have always been vitally important in the historical geography of capitalism. Since the war, however, their pivotal significance has tended to make regional change and the organization of regional divisions of labour more than ever before a direct product of sectoral dynamics, as particular industries and specific branches and firms within industrial sectors are increasingly differentiated in terms of productivity, profitability, and control over the labour force. Underlying this juxtaposition of rapid growth cycles in some sectors/branches/firms, and decline and devalorization in others, has been a far reaching technological 'fix' stimulated and sustained by state policies (especially via defence and military expenditures in the USA) and putatively aimed at achieving greater flexibility in the workplace and in the organization of the labour process.

Flexible specialization in production, in labour-management relations, in the location of productive activities, has the effect of de-rigidifying older hierarchical structures and creating at least the appearance of a significantly different order of responsibility and control. For Piore and Sabel (1984), it has defined *The New Industrial Divide*, the most significant economic and political transformation since the origins of industrial capitalism. For Claus Offe (1985), the German theorist of the state, it is part of *Disorganized Capitalism* or, as Lash and Urry (1987) call it, *The End of Organized Capitalism*, a breakdown of planned and managed systems of social power and political authority. For others, it is the essence of postfordism, and the instigator of new modes of agglomeration built around increasingly disintegrated social divisions of labour and a flexible fabric of transactions coagulating in innovative urban and regional production complexes that are located outside the centres of the old Fordist industrial landscape.

Flexible specialization, vertically disintegrated production systems, and the breakdown of rigid hierarchies have been accompanied by accelerated capital mobility to facilitate the search for sectoral super-profits (including those achieved by substantial cheapening of labour costs) anywhere in the world. The geographical search is not always successful, of course, but the aggregate effect of these sectoral restruc-turing processes has been to derigidify long-established spatial divisions of labour at virtually every geographical scale. Here is where the sectoral and spatial scenarios of contemporary restructuring converge and reverberate together, speeding up the cycles of exploitation in both the vertical and horizontal planes of uneven development.

The regional repercussions of this, perhaps unparalleled, loosening up of the landscape of capital under its new regime of 'flexible accumulation' (Harvey, 1987) have been dramatic and perplexing. Relatively stable mosaics of uneven regional development have suddenly become almost kaleidoscopic. Once highly industrialized and prosperous core regions – segments of the American manufacturing belt, north-east England and Wales, northern France, Wallonia, the Ruhr – have been experiencing accelerated economic decline and deindustrialization, while many poor peripheral regions (including some of the classical examples of regional underdevelopment) have become new centres of industrial growth and economic expansion. These 'role reversals of regions', as Mandel called them, reflect what has been the most extensive geographical decentralization and internationalization of industrial production since the origins of industrial capitalism, generating a growing list of NICs and NIRs, newly industrialized countries and subnational regions.

The role reversal of regions is in itself an oversimplification. The regional restructuring that has been taking place is much more complex and unstable. It might better be described as an accelerated regional recycling, with regions moving through several phases of development and decline in association with shifting sectoral superprofits, rounds of intensive labour disciplining, and heightened capital mobility. One of the best analysed examples of this regional recycling is New England (Harrison, 1984), now booming again after an intense internal disciplining of both capital and labour. A similar recovery may also be happening in the Scottish Lowlands and a few other older industrial regions. There is also some evidence to suggest that the industrial 'miracles' of the NIRs and NICs may be equally unstable and short-lived, making such grand descriptive metaphors as the Frostbelt–Sunbelt shift and the New International Economic Order appear increasingly exaggerated and misleading.

What has been happening can be more cautiously described as a significant but not transformative shake-up in long-established regional divisions of labour and the formation of new and still highly unstable regionalizations of national economies. Associated with this restructured regionalization has been a responsive regionalism, as various social movements and regional political coalitions react to this restructuring – to resist, to encourage, to reorganize, to demand more, to press for redirection. These multiple forms of regionalism, whether radical or reactionary, have repoliticized the regional question as a more general spatial question. No longer is regionalism rooted only in resistance to the homogenization of cultural traditions, as it was primarily in the nineteenth century. It is now part of what Goodman (1979) aptly called 'regional wars for jobs and dollars', an intensified territorial competition that stretches across the whole hierarchy of spatial locales, from the smallest locality to the scale of the world economy.

The growing importance of technological and sectoral restructuring has thus not eliminated the exploitation of geographically uneven development as a source of sustaining superprofits. Nor has it reduced the political and economic significance of the spatiality of social life. On the contrary, the contemporary period of restructuring has been accompanied by an accentuated visibility and consciousness of spatiality and spatialization, regionalization and regionalism. The instrumentality of the spatial and locational strategies of capital accumulation and social control is being revealed more clearly than at any time over the past hundred years. Simultaneously, there is also a growing realization that labour, and all other segments of society peripheralized and dominated in one way or another by capitalist development and restructuring, must seek to create spatially conscious counter-strategies at every geographical scale, in a multiplicity of locales, to compete for control over the restructuring of space. Given the effective empowerment of a neo-conservative opposition bent on burying again the exploitative instrumentality of spatial restructuring, it becomes even more urgent that all progressive social forces – feminism, the 'Greens', the peace movement, organized and disorganized labour, movements for national liberation and for radical urban and regional change – become consciously and explicitly spatial movements as well. For the left, this is the postfordist and postmodern regional challenge.

## Restructuring and the Evolution of Urban Form

Contemporary studies of urban restructuring have begun to recapitulate a historical geography that closely parallels the sequence of spatializations just described for regional development. As these retrospective glances accumulate, it becomes increasingly possible to argue that the evolution of urban form (the internal spatial structure of the capitalist city) has followed the same periodizable rhythm of crisis-induced formation and reformation that has shaped the macro-geographical landscape of capital since the beginnings of large-scale industrialization. Looking back from the present fourth modernization, the same three prior periods of accelerated restructuring and modernization stand out from the continuous restlessness. Each begins with the downturns of recession, depression, and social upheaval that have marked the end of long phases of expansive growth in the macro-political economy of

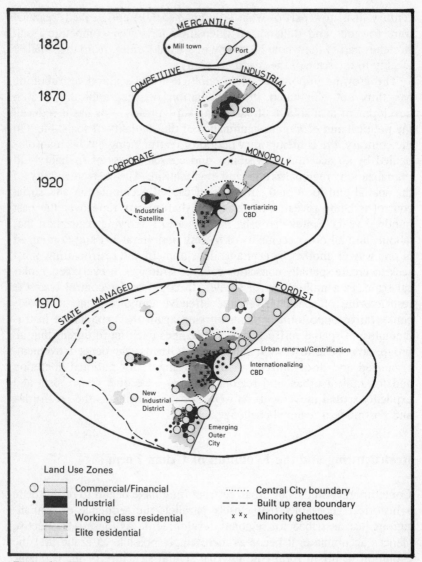

MERCANTILE

1820    • Mill town    ◯ Port

COMPETITIVE    INDUSTRIAL

1870    CBD

CORPORATE    MONOPOLY

1920    Industrial Satellite    Tertiarizing CBD

STATE MANAGED    FORDIST

1970    Urban renewal/Gentrification

Internationalizing CBD

New Industrial District

Emerging Outer City

Land Use Zones

◯ ▢ Commercial/Financial        ⋯⋯ Central City boundary
• ■ Industrial                  −−−− Built up area boundary
▨ Working class residential     x ˣx  Minority ghettoes
▨ Elite residential

*Figure 1*    The evolution of urban form: prototypes of the North American city 1820–1970

capitalist development. Each also engenders a significant recontextualization of the spatiality of social life, a different human geography.

The precise dating of these restructuring eras varies from country to country, as do the relative intensities of restructuration. But between 1830 and the mid nineteenth century, during the last three decades before the start of the twentieth century, and in the period after the Russian Revolution to the end of the Second World War, it seems increasingly clear that the internal form and social regionalization of the capitalist city experienced significant changes almost everywhere. And it is equally clear that we are now embroiled in another round of profound and perplexing urban metamorphosis.

Figure 1 presents a generalized picture of the evolution of urban form through these four restructuring periods. As with regional restructuring, the sequence of urban spatializations is cumulative, with each phase containing traces of earlier geographies, already formed urban spatial divisions of labour which do not disappear so much as become select-ively rearranged. If one feels comfortable with the geological metaphor used by Massey (1984) to describe the inter-regional spatial division of labour, the specifically urban spatializations can also be seen as 'layered' one on top of the other to reflect pronounced shifts in the geographies of investment, industrial production, collective consumption, and social struggle. The sedimentation, however, is more complex and labyrinthine than a simple layering, for each cross-section contains representations of the past as well as the contexts for the next round of restructuring.

There are other simplifications in this sequential framing, made necessary by the detail-shedding generalizations required to produce such summative mappings. The base landscape, for example, assumes no pre-existing cities projecting their pre-capitalist built environments into the urban picture. All the schematic descriptions will thus be more directly applicable to areas where urbanization and industrialization originate in conjunction with the spread of capitalism. Where there has been extensive pre-capitalist urbanization, as in Europe and Asia, the picture is much less neat and orderly, although some of the same patternings can be distinguished. The North American city probably provides the closest approximation, allowing for distortions caused by differing local physical geographies. Chicago, for example, will fit better than New York city, just as Manchester will fit better than London. The 'models' shown in Figure 1 thus serve as idealized generalizations meant to illustrate the composite of key restructuring processes which are not always found together or equally vividly in every capitalist city.

The sequence begins with the '*Mercantile City*' set in a context of petty commodity production, international trade, and limited industrial-ization. Its port city focus and sprinkling of small industrial milltowns in a still largely agricultural landscape most closely characterizes the urban situation in the United States before the 1840s. In the exemplifying American case, commercial capital was concentrated in small, dense, mainly coastal settlements occupied by artisans, shopkeepers, farmers, administrators, and traders. New York, Philadelphia, Baltimore, and

Boston were the primary centres in the North. Savannah, Charleston,
Mobile, and New Orleans, much smaller than the big four of the North,
served the southern economy and the slave-based cotton production
that provided the major national export commodity.

Eighty per cent of the urban population was self-employed and urban
life revolved primarily around the activities of the petty bourgeoisie and the
nodality of the port focus. Capitalist industrialization, as was true of the
early phases of the industrial revolution in Europe, began primarily in
the countryside at energy sites connected to the commercial ports by
canal, river transport and, increasingly, by rail. Although social struggles
tended to be centred around agrarian issues and rural landownership
during most of the post-Revolutionary period, rising inequalities of
wealth within the mercantile cities became increasingly visible and these
inequalities frequently became the source of conflict and social unrest.

David Gordon, summarizing the rich historical literature on early
American urbanization, provides a spatially insightful description of the
'contradictions of the Commercial City':

> Commercial accumulation tended to generate uneven development among
> buyers and sellers.... Because different socioeconomic groups were living and
> working closely together in the Commercial Cities, these spreading inequali-
> ties became more and more physically evident.... It appears that this evidence
> of inequality generated popular protest against it. As inequalities reached their
> peak during the 1820s and 1830s, popular protests also seemed to intensify.
> Most of these protests focused on demands for more equality. Because these
> protests frequently had political effect, they tended to limit opportunities for
> further commercial accumulation. (1978, 36)

Gordon adds that this 'dialectic of uneven development and popular
protest' demonstrated the 'fundamentally *spatial* aspect to the contradictions of the commercial path to capital accumulation'.

> Because the Commercial City retained the precapitalist transparencies of
> immediate, intimate, and integrated social relationships, commercial capitalist
> profits could not be masked. The quest for such a disguise – the urgent need
> for which was so dramatically witnessed in the streets of the Commercial City
> – played a central role in prompting a turn to a new and ultimately more
> opaque mode of capital accumulation. (Ibid.)

Similar conditions brought popular protest into the streets of European
and other cities in 1830 and 1848, the most explosive moments of the
first major phase of urban restructuring.

The period from the 1840s to the early 1870s was a time of extraordinary industrial and urban growth in Europe and in North America.

International trade simultaneously expanded at an unprecedented rate. The 'age of capital', as Hobsbawm called it, had superseded the 'age of revolution' and the pace of capitalist spatialization accelerated at every geographical scale after the significant restructuring of capitalism that culminated in the global crises of 1848–49. The punctuation points of this spatialization were new kinds of cities and hierarchical city-systems which added to the growing traditional functions of social control, commercial accumulation, and political administration an unprecedented, scale-expanding agglomeration of industrial production.

Nodality was particularly important for industrial capital. It brought about much greater returns with each increase than it did for commerce or agriculture and once freed from certain physical and historical constraints the characteristic nodality of human locales exploded to create the distinctive form of the classic '*Competitive Industrial Capitalist City*'. Never before was production so geographically concentrated, so locationally centralized, so densely agglomerated. In the eastern United States, urban industrialization swallowed up the small mercantile cities, either eliminating them entirely or preserving some remembrances in vestigial 'old towns' that have occasionally survived to the present (or have been recreated as simulacra, exact copies for which the original no longer exists). The long-established pre-industrial cities of Europe were much more difficult to disintegrate and digest, but even there the centralization of industrial production began to disassemble the urban cores to accommodate the expansion of the three distinctive elements of competitive industrial capitalism: the industrial factories and associated producer services, the industrial bourgeoisie, and the new urban proletariat, the proximate industrial working class that made the whole system work.

The intensification of land use in the urban centre redefined the form of the city and instigated a remarkable – and more opaque – social and spatial ordering of urban life. Accommodative technologies of transport and building (for example, the railway and the lift) accelerated this intensification and its associated wellspring of agglomeration economies. Rippling out from the Central Business District and employment nucleus was a zoned built environment of residentiary rings and radial sectors gridded to contain the attenuated daily journeys to work (for the urban proletariat) and the daily journeys to control workers (for the industrial bourgeoisie). The zonation was largely a matter of class, as the antagonistic social structure of competitive industrial capitalism became spatialized in segregated and socially homogeneous urban compartments and enclosures.

The internal regionalization was much more repetitive and neatly mappable than in pre-industrial cities, but the flagrant regularity was not often seen by scholarly observers at the time. The social and political significance of the spatial organization of society was conveniently fading from view in nineteenth-century social science and only a few perspicacious onlookers, such as Engels in his writings on Manchester, were able to see the increasingly hidden instrumentality of this new urban geography. In its broad outlines of concentricities and wedges, centres and peripheries, as well as in its more intricate web of cells, ranks, enclosures, and partitionings, this was a disciplinary spatialization designed not by some conspiracy of capitalist architects but artfully designed nonetheless.[2]

The Chicago School of urban ecologists uncovered the broad geometry of this particular urban regionalization but buried its powerful instrumentality under an obfuscating ideology of naturalism and/or cultural relativism. Later this instrumentality would be buried even deeper, as the logic of neo-classical economics took hold of urban theory and made it a disciplining space of its own. But well before these developments, the form of the centralized industrial capitalist city had changed dramatically, even in paradigmatic Chicago. The explosion of the Paris Commune in 1871, the financial panic of 1873, the associated loss of 300,000 jobs in the United States, a series of mass evictions and rent riots in New York City, and other shocks to the system had signalled the end of the post-1848 boom and the start of another prolonged period of capitalist restructuring which would last almost until the end of the century.

During this restructuring period, capitalism was markedly intensified through the increasing concentration and centralization of capital in corporate monopolies; and extensified through the expansive internationalization of finance capital in an emerging age of imperialism. The state began to enter more deeply into the economy, especially through fiscal and monetary management and the initiation of urban economic planning. As this was occurring, there were also vigorous working-class protests and strikes, led by growing unions and socialist/labour parties in virtually all the industrialized countries.

Labour unrest in the United States, often instigated by radicalized European immigrants, was first centred on small mining and railway towns, but after 1885 it spread into the centres of the largest cities and

---

2. Harvey's rich essay on the transformation of Paris, 1850–70 (Harvey, 1985b) explores perhaps the most paradigmatic example of the disciplinary power contained in this instrumental urban spatialization, especially with its extraordinary Haussmanesque elaborations. Foucault begins his explorations much earlier and is more responsible than anyone else for opening our eyes to the instrumentality of space/power/knowledge in the urban context.

exploded with particular vehemence in the three major industrial centres, New York, Philadelphia, and Chicago (Gordon, 1984). The efficient geographical centralization of factories and working-class communities which characterized the classic industrial capitalist city seemed to be breeding a strengthened working-class consciousness and militancy, paralleling on a much larger scale the earlier experience of company towns. Not only had capitalist accumulation slowed down, but the disciplinary spatiality of urban life had been weakened. Even Hauss-man's Paris had been taken over, if only briefly, by the Communards.

The crisis-induced need to restore conditions of profitable capital accumulation and labour control lay behind the intensification and extensification of capitalism that developed through the late nineteenth century. After the world-wide depression of the 1890s, capitalism rebounded with new vigour in the industrialized countries and accompanying this *fin de siècle* turnaround was a selective restructuration of the capitalist city. By the 1920s a new geography had consolidated, more clearly in some cities, to be sure, than in others. I will again use the American city as the archetypal example.

Greater corporate centralization, an increased segmentation of the labour force into monopoly and competitive sectors, new production technologies, and the separation of management and production functions reorganized the spatial division of labour in capitalist urbanization. In the new '*Corporate-Monopoly Capitalist City*', industrial production became less concentrated around the city centre, as factories spread into the formerly residential inner rings and, even more against the old pattern, into satellite industrial centres such as Gary, Indiana, and East St Louis, Illinois. As a result, the old urban cores became increasingly tertiarized, replacing lost industries with an expanding number of corporate headquarters (by 1929, more than half the largest corporate headquarters were located in New York and Chicago alone), government offices, financial institutions, and supportive service and surveillance activities.

Rings of working-class residences and racial and ethnic ghetto enclaves continued to serve a still dominant urban core, and in almost every large American city there remained at least one protected residential wedge stretching from centre to periphery where the highest income inhabitants had their homes. But the cityscape sprawled much further outwards as managers, supervisors, and professionals joined the old industrial bourgeoisie in a surge of suburbanization that broke out of the earlier administrative boundaries of the city. The areal multiplication of separate incorporated municipalities replaced annexation as the major pattern of urban territorial expansion, creating a degree of metropolitan political fragmentation never approached in the past. The urban landscape was not only stretched over a much larger area, it was broken into many more pieces.

This fragmented, polycentric, and much more complex urban region-alization assisted industrial capital in escaping from agglomerated working-class militancy. Employers could more easily move away from organized union pressures,[3] the workforce became more segmented and residentially segregated, new assembly-line and related technologies provided less agglomerated ways to capture the positive externalities of industrial production, and a more beneficent local and national state could be tapped more easily for substantial subsidies. The geography was not as neatly ordered as it had been but, especially with the motor car and other forms of mass transit, the urban possibilities for capital accumulation and labour co-optation were significantly expanded. By the time the boom was over, the American city had almost completed its captivating proletarianization. Not much more than a century earlier, the urban population was around 80 per cent self-employed. By 1940, almost 80 per cent were wage and salary workers.

The after-effects of the Russian Revolution and then the onset of the Great Depression shattered the complacencies of the early twentieth-century boom and led to still another round of urban restructuring. Rather than creating any great new turnarounds and reversals, however, this third restructuring seemed to amplify further the same processes that characterized the second modernization: increasing centralization, concentration, and internationalization of corporate capital; increasing segmentation of labour based on a changing organization of the pro-duction process; increasing urban political fragmentation and disagglo-meration of working-class communities; and an even greater role for the state in both fostering accumulation and maintaining legitimized labour discipline.

Through Keynesian policies of demand stimulation, monetary and fiscal controls, economic planning, state-directed investments in indus-try, and social welfare programmes (aimed specifically at assuaging working-class pressures and dampening down social unrest), the state intervened more directly and deeply into the production process than ever before. A '*State-Managed Urban System*' began to take shape from the devalorization and restructuring of the Corporate City in the Depression years. After wartime devastation opened still more room for

---

3. Gordon (1984, 41) cites the president of a contracting firm in Chicago: '... all these controversies and strikes that we have had here for some years ... have prevented outsiders from coming in here and investing their capital ... It has discouraged capital at home ... It has drawn the manufacturers away from the city, because they are afraid their men will get into trouble and get into strikes ... The result is, all around Chicago for forty or fifty miles, the smaller towns are getting these manufacturing plants.'

accelerated accumulation and super-profit making, capitalism boomed again in another guise. A different urban landscape consolidated in the wake of this expansion.

Suburbanization was markedly accelerated after the Second World War. With substantial state support and encouragement, sizeable portions of the working class, blue as well as white collar, settled into suburban tracts and privatized enclosures. Expansive metropolitanization, accompanied by an even greater fragmentation of political jurisdictions and a quickened decentralization (not only of industrial plants but also of corporate headquarters, retail, and other services) contributed further to the selective abandonment of the inner urban core. The old centres of the urban landscape were left with a residual mix of competitive sector firms, older industries, some luxury shops and hotels, key agencies of the state and finance capital, remaining corporate headquarters, and a bloated irregular workforce comprised primarily of minorities and the poorest segments of the metropolitan population – a geographically concentrated and subservient reserve army of labour.[4]

The major devaluation of inner city assets associated with the consolidation of the State-Managed Urban System was linked from the start with persistent and usually state-assisted efforts at downtown 'revitalization' – through urban renewal, gentrification, and changes in landownership and regulatory patterns aimed at maintaining a substantial corporate (and managerial) presence. The delicate balance between deterioration and renewal varied from city to city and changed significantly over time, creating an uneasy situation which required massive expenditure by the state in the form of welfare payments for both the poor and the wealthy who remained – added to the subsidization to suburbanize those who moved out. The tensions surrounding this uneasy mix – so redolent of the knife-edge spatial problematic described by Harvey – permeated urban (and to a significant extent, national) politics throughout the post-war period, and not only in the United States.

The ability to sustain this seemingly unstable spatialization and the co-ordinated spatial planning it required derived largely from the expansionary boom that began in the USA during the war and in Europe and Japan some time after. Again, the state was in the forefront. Automobile manufacturing (by state-owned companies in many countries), air transport (most often state-run), the oil industry (heavily subsidized to fuel expanded physical mobility), housing construction (fostered by govern-

---

4. There were similar processes operating in European cities, but suburbanization, metropolitan political fragmentation, and the abandonment of the inner city (except via wartime destruction) tended to be much less intense.

ment programmes for mortgages and loans), and the production of consumer durables (televisions, washing machines, and other commodities allowing for increased privatization of formerly more collectively organized consumption) led the economic expansion and contributed to major changes in urban form, especially through the concurrent suburbanization process (Walker, 1981).

More than ever before, the social and spatial relations organizing production and reproduction and the conflicts and struggles arising from these relations came to be channelled, absorbed, and managed by the state. Some have called this particular regime of accumulation and mode of regulation 'state monopoly capitalism'. Poulantzas often used 'authoritarian state capitalism' and Lefebvre once toyed with calling it a *système étatique*, virtually a state mode of production. For Mandel, 'Late Capitalism' was sufficient. Other soubriquets attached accumulation and regulation to Fordism (Lipietz, 1986) to re-emphasize their rootedness in the industrial labour process. Whatever its most appropriate label, this 'different' capitalism produced a 'different' urban spatialization, a provisional urban spatial fix filled again with new possibilities as well as the seeds of its own recreative destruction. This altered, but still contradiction-filled urban landscape, appearing most vividly in the 1950s and 1960s, was the backdrop for the development of a Marxist urban political economy and urban sociology, a Marxism which for the first time became centred on the specificities of urbanization and spatial change. But before the landscape could be effectively understood, it too began to change.

The onset of the current period of urban restructuring can be traced into the stretched and tensely matted fabric of the state-managed and spatially planned urban systems – the *villes sauvages* Castells called them – that began to be torn apart in the 1960s. The contagious inner-city riots sparked by those groups who benefited least from post-war boom, the urban insurrections in France and Italy in 1968 and 1969, and a multiplying series of urban-based social movements again stirred visions of a specifically urban revolution and signalled the end of post-war business-as-usual. The so-called 'urban crisis', however, was part of a much larger crisis of the state and the whole system of management, planning, welfare, and ideological legitimization that had dragged capitalism out of the Great Depression and propelled it through the prolonged post-war expansion. By the 1960s, the expensive Keynesian welfare policies had become increasingly difficult to sustain, while the financial pressures accompanying the huge expansion of credit that fuelled the boom now started fuelling inflation instead. The global recession of 1973–75 followed a chain of shocks to the system and helped to trigger another concerted round of restructuring.

As before, the very same social and spatial structures of accumulation that had facilitated expansion became the arenas of economic stress and decline. The post-war 'productivity deals' that brought major wage and welfare benefits to organized labour in return for productive peace in the workplace; the demand-driving suburbanization process; the delicately balanced devaluation/renewal of inner cities; the system of state regulation and management; the rapid expansion of government employment, shifted from being part of the solution to being part of the problem. Exacerbated by a rise in petroleum prices and energy costs and successful challenges to American global military hegemony, another political and economic crisis shattered the confidence and optimism that marked the 1950s and 1960s in the USA and other core capitalist countries. The form of the crisis was both old and new. In many ways, it involved the classic problems of overproduction/underconsumption, but complicating a classic interpretation was a reproduction crisis arising primarily from a fiscal crisis of the state and its squeezing effect on both capital accumulation and the ability to maintain effective means of labour discipline and social control. As a result, the need to reconfigure the spatial landscape of capital took on an even more crucial urgency. As this contemporary restructuring advances, it is disassembling not only the urban fabric but the framework of critical interpretation of capitalist development as well.

## Some Contemporary Conclusions and Continuities

It is still too early to make any unequivocal statements about the present phase of social and spatial restructuring. The outcome is still open, a new dye has not been rigidly cast over the restless landscape, and thus the capability to look back on a *fait accompli* is not yet available. It should be remembered that the most acute observers of the three past periods of prolonged restructuring (amongst whom I would include Marx for the first, Lenin for the second, and Mandel for the third) had the great advantage of hindsight, the opportunity to interpret a restructuring that had successfully restored the expansiveness of capitalist accumulation and had begun to consolidate its representative spatialization. A new upswing, however, has not yet begun. We must be satisfied today primarily with the tentative identification of trends and tendencies that appear to be taking hold with particular force, recognizing again that the recovery of capitalism through restructuring is not mechanical or guaranteed, that all that seems solid today may melt – or explode – into the air tomorrow.

If we succeed in being informed by past periods of restructuring, however, some provisional expectations can be outlined. The contemporary period must be seen as another crisis-generated attempt by capitalism to restore the key conditions for its survival: the opportunity for gaining superprofits from the juxtaposition of development and underdevelopment in the hierarchy of regionalized locales and amongst various productive sectors, branches, and firms. Central to the resurrection of expansive superprofits is, as usual, the institution of invigorated means of labour discipline and social control, for the sustaining logic of capitalist accumulation breeds a competitive political and economic struggle and never proceeds without friction and resistance.

This search for superprofits and enhanced social control can be divided into two broad strategic categories of intensification (the deepening of the division of labour, the generation of new consumption needs, the incorporation of new spheres into capitalist production, the greater concentration and centralization of capital, increased legitimization of the dominant ideology, the weakening of labour organization and militancy) and extensification (a 'widening' of the division of labour, the opening of new markets, geographical expansion to tap sources of cheap labour and raw materials, increasing the scope of exploitation of geographically uneven development through value transfers and unequal exchange). Thus there are many specific paths to choose from. To choose only one as the dominant path is to be foolishly inflexible and politically naive. Furthermore, the particular mix of strategies (and counter-strategies) is itself unevenly developed and never predetermined in its effectiveness.

The same claims can be made for the spatialization which accompanies this restless search for profits and discipline. It has a broad patterning to it, but is highly differentiated and unevenly developed by its very nature, taking a variety of specific forms, not all of which can be seen as 'functional' for the logic of capital or inherently antagonistic to the demands of labour. The sequential picture of the evolution of urban form and uneven regional development that I have presented here is no more than a 'thin' description which has sifted out the particularities and complexities to highlight those instrumental textures of restructuring that can be gleaned from spatial hindsight. Extending the picture to the presently unfolding urban and regional restructuring is more difficult and will require a much more substantial empirical understanding and theoretical adaptation than has yet been achieved.

Nevertheless, looking back from the present moment to the events of the past two decades provides at least a tentative basis for identifying a series of indicative trends characterizing the contemporary restructuring process. Each of these trends has a spatializing impact that has become increasingly evident in the 1980s.

1. One prevailing trend has been the increasing centralization and concentration of capital ownership, typified by the formation of huge corporate conglomerates combining diversified industrial production, finance, real estate, information processing, entertainment and other service activities. This conglomeration process goes several steps further than the horizontal mergers of the late nineteenth century and the later vertical mergers and state-managed monopolies that were so central to the rise of Fordism. Formal management structures are often less centrally controlled and more flexible, while the core production processes have increasingly been broken into separate segments operating, unlike the integrated Fordist assembly line, at many different locations. The multiplication of branches adds further flexibility through parallel production, while more extensive subcontracting expands the vital transactions of the conglomerate even further beyond the bounds of ownership.

2. Added to the corporate conglomeration of ownership has been a more technologically-based integration of diversified industrial, research, and service activities that similarly reallocates capital and labour into sprawling spatial systems of production linking centres of administrative power over capital investment to a constellation of parallel branches, subsidiaries, subcontracting firms, and specialized public and private services. More than ever before, the spatial scope of these production systems has become global, but they also have a powerful urbanization effect through the local agglomeration of new territorial industrial complexes (usually located outside the old centres of Fordist industry). Here again there seems to be a paradoxical pairing of deconcentration and reconcentration in the geographical landscape.

3. Linked to increased capital concentration and oligopoly has been a more pronounced internationalization and global involvement of productive and finance capital, sustained by new arrangements for credit and liquidity organized on a world scale. This transnational or global capital is able to explore and exploit commodity, financial, consumer, and labour markets all over the world with fewer territorial constraints (especially from direct state control) than ever before. As a result, purely domestic capital has been playing a decreasing role in the local and national economies of the advanced industrial countries as these economies increasingly internationalize.

4. The weakening of local controls and state regulation over an increasingly 'footloose' and mobile capital has contributed to an extraordinary global restructuring of industrial production. Large-scale capitalist

industrialization has been occurring in a series of peripheral countries and regions for the first time, while many core countries have been experiencing an extensive regional industrial decline. This combination of deindustrialization and reindustrialization has shattered long-standing global definitions of core and periphery, First-Second-Third Worlds, and created the tentative outlines of a different, if not an entirely new, international division of labour.

5. In the USA and elsewhere, the accelerated geographical mobility of industrial and industry-related capital has triggered and intensified territorial competition among government units for new investments (and for maintaining existing firms in place). These 'regional wars for jobs and dollars' (Goodman, 1979) absorb increasing amounts of public funds and often dominate the urban and regional planning process (at the expense of local social services and welfare). As capital increasingly co-operates, communities increasingly compete, another old paradox that is becoming particularly intensified in the present period.

6. Paralleling what has been happening at the global scale, the regional division of labour within countries has been changing more dramatically than it has over the past hundred years. Regions containing the manufacturing sectors that led the Fordist postwar boom (motor cars, steel, construction, civilian aircraft, consumer durables) are being disciplined and 'rationalized' through a varying mix of capital flight and plant closures, the introduction of new labour-saving technology, and more direct attacks on organized labour (deunionization, labour givebacks, constraints on collective bargaining). A selective reindustrialization based primarily on advanced technologies of production and centred on less-unionized sectors is simultaneously either arresting the decline in a few of the more successfully rationalized regions (for example, New England) or focusing industrial expansion on new territorial industrial complexes (typically on the periphery of major metropolitan areas).

7. Accompanying these processes are major changes in the structure of urban labour markets. Deeper segmentation and fragmentation is occurring, with a more pronounced polarization of occupations between high pay/high skill and low pay/low skill workers, and an increasingly specialized residential segregation based on occupation, race, ethnicity, immigrant status, income, lifestyle, and other employment related variables. An overall decrease in the relative proportion of manufacturing employment (due mainly to declining employment in the older, more unionized heavy industries), accompanied by a rapid increase in lower wage tertiary employment, tends to produce significantly reduced

(if not negative) rates of growth in wage levels and real income for workers and curtailed increases in productivity levels in the national economy wherever this shifting sectoral structure is most pronounced.

8. Job growth tends to be concentrated in those sectors which can most easily avail themselves of comparatively cheap, weakly organized, and easily manipulable labour pools and which are thus better able to compete within an international market (or obtain significant protection against international competition from the local or national state). The leading job growth sectors are thus both high and low technology based, and draw upon a mix of skilled technicians, part-time workers, immigrants and women. This creates a squeeze in the middle of the labour market, with a small bulging at the top and an even greater bulging at the bottom (especially if one includes the burgeoning informal economy as well). Only in the United States, however, among all the advanced industrial countries at least, has this dramatic employment restructuring been associated with substantial aggregate job growth.[5]

These and other prevailing restructuring processes have injected a peculiar equivocalness into the changing geographical landscape, a combination of opposites that defies simple categorical generalization. Never before has the spatiality of the industrial capitalist city or the mosaic of uneven regional development become so kaleidoscopic, so loosened from its nineteenth-century moorings, so filled with unsettling contrariety. On the one hand, there is significant urban deindustrialization emptying the old nodal concentrations not only to the suburban rings, a pattern which began as far back as the late nineteenth century, but much further afield – into small non-metropolitan towns and 'greenfield' sites or beyond, to the NICs and NIRs. On the other hand, a new kind of industrial base is being established in the major metropolitan regions, with an 'urbanization effect' that is almost oblivious to the locational advantages embedded in the former urban–industrial landscape (Scott, 1983a, 1986). To speak of the 'post-industrial' city is thus,

---

5. The nature of this 'Great American Job Machine', as some have begun to call it, is still difficult to grasp. Suggestive, however, are two recent lists of the ten fastest growing occupations, published by the Institute for Research on Educational Finance and Governance at Stanford University. In terms of absolute numbers, the top ten occupations are building custodian, cashier, secretary, general office clerk, sales clerk, professional nurse, waiter/waitress, kindergarten/elementary school teacher, truck driver, and nurse's aide/orderly. The largest percentage increase expected over the next decade (reflecting the past decade) are for computer service technician, legal assistant, computer systems analyst, computer programmer, office machine repairer, physical therapy assistant, electrical engineer, civil engineering technician, personal computer equipment operator, and computer operator.

at best, a half-truth and at worst a baffling misinterpretation of contemporary urban and regional dynamics, for industrialization remains the primary propulsive force in development everywhere in the contemporary world.

Growing, in large part, out of this combination of deindustrialization and reindustrialization is an equally paradoxical internal restructuring of metropolitan regions, marked by both a decentering and recentering of urban nodalities. Sprawling suburbanization/metropolitanization continues but it no longer seems as unambiguously associated with the decline of the downtown centres. Carefully orchestrated downtown 'renaissance' is occurring in both booming and declining metropolitan regions. At the same time, what some have called 'outer cities', rather amorphous agglomerations that defy conventional definitions of urban–suburban–exurban, are forming new concentrations within the metropolitan fabric and provoking a spray of neologisms which try to capture their distinctiveness: technopolis, technoburb, urban village, metroplex, silicon landscape.

The internationalization process has created another set of paradoxes, for it involves both a reaching out from the urban to the global and a reaching in from the global to the urban locale. This has given new meaning to the notion of the 'world city' as an urban condensation of the restructured international division of labour (Friedmann and Wolff, 1982). More than ever before, the macro-political economy of the world is becoming contextualized and reproduced in the city. First World cities are being filled with Third World populations that, in some cases, are now the majority. While these combinatorial world cities increasingly stretch out to shape the international economy in a form of global spatial planning, they also increasingly incorporate internally the political and economic tensions and battlegrounds of international relations.

Neither conventional urban theory nor the Marxist urban political economy that consolidated in the 1970s has been able to make theoretical and political sense of this enigmatic contemporary urban restructuring. Whereas the former tends to overspecify the urban, making the assertion of urbanism take the place of explanation, the latter has tended, for the most part, to underspecify the urban, passing abstractly over its causal power and its integral positioning within the historical geography of capitalism. Both have tended to overemphasize consumption issues and neglect the urbanization effects of industrial production, a narrowing which may have been politically appropriate to the 1960s but is now too short-sighted to contend effectively with contemporary restructuring processes.

In many ways, the same can be said for the neo-Marxist international political economy that evolved in tandem with the urban. It, too, tended to oversimplify the complexities of capitalist production and labour processes, or to assume that historical materialism had already solved all its riddles. Much was accomplished in exploring the multiple circuits of capital shaping the world system and in retracing their historical origins and geographical development. But the prevailing perspectives were caught short by the dramatic shifts in the international division of labour brought about by an essentially unexpected and world-wide industrial restructuring.

At present, the relatively new field of regional political economy and a reinvigorated and reoriented regional industrial geography seem to be the most insightful and innovative arenas for analysing the macro-, meso-, and micro-political economies of restructuring. Both can be called flexible specializations, for they are less concerned with old boundaries and disciplinary constraints and are thus more open to timely adaptation to meet new demands and challenges. The regional perspective facilitates the synthesis of the urban and the global while remaining cognizant of the powerful mediating role of the national state even as this role dwindles somewhat in the current era. The mutually responsive interplay of regionalization and regionalism provides a particularly insightful window on to the dynamics of spatialization and geographically uneven development, gives greater depth and political meaning to the notion of spatial divisions of labour, and abounds with useful connections to the revamped social ontologies discussed earlier. Just as important, its openness and flexibility, its inclination to try new combinations of ideas rather than fall back to old categorical dualities, makes *critical regional studies* the most likely point of confluence for the three streams of contemporary restructuring. Here is where our understanding of postfordism, postmodernism, and a post-historicist critical social theory may most bountifully take place.

# 8

# It All Comes Together in Los Angeles

I should be very much pleased if you could find me something good (meaty) on economic conditions in *California*.... California is very important for me because nowhere else has the upheaval most shamelessly caused by capitalist centralization taken place with such speed.

(Letter from Karl Marx to Friedrich Sorge, 1880)

There it is! Take it!

(Dedication speech by William Mulholland
at the opening of the Los Angeles Aqueduct, 1913)

Marx's premonitory curiosity about California was piqued by the extraordinary events following the gold discoveries of 1848. Out of practically nowhere, a formidable capitalist presence emerged along the Pacific Ocean rim of the New World, beginning a Californian tilt to the global space economy of capitalism that would continue for the next century and a half. California gold significantly fuelled the recovery and expansion of industrial capitalism after the age of revolution, helped prime the pump for the territorial consolidation and rapid urban industrialization of the United States, and deposited in the San Francisco Bay region one of the late nineteenth century's most dynamic centres of accumulation. But the process, once begun, did not end there.

Relatively unseen in 1880 was the onset of another, more local, tilting that would sustain the Californianization of capitalism through the twentieth century. The rise of Southern California, the region centred on the city of Los Angeles, has confirmed the prescient intuition of Marx. Since 1900, there may be no other place where the upheavals associated with capitalist centralization have developed more rapidly or

*190*

shamelessly. What Northern California was to the last half of the nineteenth century, Southern California has been much more to the twentieth.

Were we able to create decennial global maps of the regional generation of superprofits starting in the 1920s, the Los Angeles area would almost surely be among the peak points in virtually every decade. Green and black gold – agriculture and oil – sustained the earliest expansion, with Los Angeles County leading the country in both sectors of production for many decades. From the 1930s, however, industry took the lead and, in decade after decade up to the present, Southern California was never substantially surpassed by any other comparable American region in the net addition of manufacturing employment. For most of the twentieth century, Los Angeles has been amongst the most propulsive and superprofitable industrial growth poles in the world economy, consistently localizing the leading industrial sectors of the moment. Today, the regional economic product of Southern California is larger than the gross national product of all but ten countries.[1]

What better place can there be to illustrate and synthesize the dynamics of capitalist spatialization? In so many ways, Los Angeles is the place where 'it all comes together', to borrow the immodest slogan of the *Los Angeles Times*. Being more inventive, one might call the sprawling urban region defined by a sixty-mile (100 kilometre) circle around the centre of the City of Los Angeles a *prototopos*, a paradigmatic place; or, pushing inventiveness still further, a *mesocosm*, an ordered world in which the micro and the macro, the idiographic and the nomothetic, the concrete and the abstract, can be seen simultaneously in an articulated and interactive combination. But let us back off from these presumptive characterizations for the moment to aim more modestly at theoretically informed regional description, at an empirically based case study in the historical geography of urban and regional restructuring.

## The Contemporary Setting

The Los Angeles urban region (covering the five counties of Los Angeles, Orange, Ventura, Riverside, and San Bernardino) is today one of the largest industrial metropolises in the world, having recently passed

---

1. For the whole state of California, the figures are even more astounding. In 1987, a proud advertisement in *Business Week* (paid for by the State) announced that California had surpassed Britain and Italy to become the world's sixth largest economy, with a total output of goods and services worth $550 billion.

Greater New York in manufacturing employment and total industrial production. Moreover, since the late 1960s, it has experienced a concentration of industry, employment growth, and financial investment that may be unparalleled in any advanced industrial country. Between 1970 and 1980, when the entire USA had a net addition of less than a million manufacturing jobs and New York lost nearly 330,000, the Los Angeles region added 225,800. In the same decade, the total population grew by 1,300,000 but the number of non-agricultural wage and salary workers increased by 1,315,000, making the region by far the world's largest job machine, a position it has continued to hold in the 1980s.[2]

The regional job machine has churned most actively at two levels. Employment and production in high technology industries have expanded to make Greater Los Angeles perhaps the world's largest 'technopolis' with more engineers, scientists, mathematicians, technical specialists – and more high security cleared workers – than any other urban region. An even greater expansion in low-paying service and manufacturing jobs (with a booming garment industry leading the way) and an explosion in part-time and 'contingent' work (flexibly organized to meet changing labour demands) has ballooned the bottom of the labour market to absorb most of the nearly two million new job-seekers (mainly immigrants and women) entering the market over the past twenty years.

There have been other formidable agglomerations. A growing flow of finance, banking, and both corporate and public management, control, and decision-making functions have made Los Angeles the financial hub of the Western USA and (with Tokyo) the 'capital of capital' in the Pacific Rim. The region also contains the largest node of government employees in any American city outside Washington, DC; and the twin ports of San Pedro and Long Beach are now amongst the largest and fastest growing in the world in terms of imports and exports. Today they handle nearly half of the trans-Pacific trade of North America. And lest it be forgotten, the Los Angeles region has, since the Korean war at least, been the primate region in the country in the receipt of defence contracts for weapons research and development, the foremost arsenal of America.

Juxtaposed against these indicators of rapid aggregate growth in the regional economy, however, are equally startling indicators of decline and economic displacement: extensive job loss and factory closures in the most unionized sectors of blue-collar industry and a steep decline in the membership of industrial labour unions; deepening poverty and

---

2. The Houston region ranked second to Los Angeles in job generation during the 1970s. Its total was almost 700,000, or barely half that of the Greater Los Angeles area.

unemployment in those neighbourhoods left behind to fend for themselves in a growing informal or underground economy; the multiplication of industrial sweatshops reminiscent of the nineteenth century; the intensification of residential segregation in what has always been a highly segregated city-region; unusually high rates of violent crime, gang murders, and drug use, as well as the largest urban prison population in the country. A particularly acute housing crisis has been boiling for many years, reversing the long trend toward increasing homeownership and inducing an extraordinary array of disparate housing strategies. Perhaps as many as 250,000 people in Los Angeles County are living in transformed garages and backyard buildings, with half as many crowded into motel and hotel rooms hoping to save enough to pay the required security deposits on more stable but out of reach rental accommodations. Many are forced to 'hotbed', taking turns sleeping on never-empty mattresses, while others find accommodation in cinemas which obligingly reduce their charges after midnight. Those even less fortunate live on the streets and under the freeways, in cardboard boxes and makeshift tents, pooling together to form the largest homeless population in the United States – another 'first' for Los Angeles.

Seemingly paradoxical but functionally interdependent juxtapositions are the epitomizing features of contemporary Los Angeles. Coming together here are especially vivid exemplifications of many different processes and patterns associated with the societal restructuring of the late twentieth century. The particular combinations are unique, but condensed within them are more general expressions and reflections. One can find in Los Angeles not only the high technology industrial complexes of the Silicon Valley and the erratic sunbelt economy of Houston, but also the far-reaching industrial decline and bankrupt urban neighbourhoods of rust-belted Detroit or Cleveland. There is a Boston in Los Angeles, a Lower Manhattan and a South Bronx, a São Paulo and a Singapore. There may be no other comparable urban region which presents so vividly such a composite assemblage and articulation of urban restructuring processes. Los Angeles seems to be conjugating the recent history of capitalist urbanization in virtually all its inflectional forms.

## A Brief Historical Glance

Los Angeles never fully experienced the intensive geographical central-ization of production that characterized the nineteenth-century indus-trial capitalist city and shaped the early expansion of most large

American cities east of the Rockies (and San Francisco to the north). Although founded in 1781, the city of Los Angeles remained a small peripheral outpost until a century later, when the prevailing urbanization process had become more decentralized, extensive residential suburbanization had begun, and clusters of separately incorporated municipalities started to rim the central metropolitan city. The rapid population growth which occurred between 1880 and 1920, when Los Angeles County expanded from 35,000 to nearly a million, was thus shaped primarily by the social and spatial relations of the Corporate City.

But there was never any doubt about where the 'centre' was located. Government, financial, and commercial activities have always been concentrated in the downtown core of the City of Los Angeles, the beacon of social control and administration for more than two hundred years. There was also a sizeable inner industrial zone just to the south which would eventually expand, almost continuously, to the port complex of San Pedro (annexed to the city in 1909) and Long Beach (incorporated first in 1888 and, since 1920, the second largest city in the region). The prevailing pattern of residential and industrial location, however, was already polynucleated and decentralized, with relatively low densities in the sprawling built-up area.

'Black gold suburbs' (Viehe, 1981) pockmarked Los Angeles County outside the three largest centres of Los Angeles, Pasadena, and Long Beach (itself perhaps the biggest of these petroleum-linked urbanizations), and spread into the counties of Orange and San Bernardino. The black gold suburbs included Whittier, Huntington Beach, Norwalk, El Segundo (literally defining Standard Oil's second major refinery), and many other locations which were swallowed up by annexation to the central city. An efficient and extraordinarily extensive rail network of mass transit – the largest in the country at the time – connected these scattered centres and fostered, with each connection, a pattern of intensely competitive locality boosterism and territorial fragmentation.

From a landlocked 43 square miles at the beginning of the century, the city of Los Angeles grew to 362 square miles in 1920, and 442 in 1930 by reaching out annexing tentacles first to the port and then, through its virtual monopoly on water, to Hollywood, Venice, Watts, and vast stretches of the San Fernando Valley. William Mulholland's laconic dedication speech at the opening of the Los Angeles Aqueduct in 1913 – 'There it is! Take it!' – was clearly being taken seriously. But even with this aggressive expansion of the city, the other incorporated areas of the county soon began to grow more rapidly in population. After 1920, the city never again experienced an intercensus population growth rate greater than the rest of the county.

From 1920–40, covering the years of the Great Depression, Los

Angeles County added nearly two million inhabitants, roughly evenly divided between city and suburbs (although by this time the distinction had already become exceedingly blurred, what with most of the city being as suburban as the so-called suburbs). Urban growth was based mainly on petroleum extraction and refining, agriculture, the motion picture industry, unusually aggressive forms of land speculation and real-estate development, and a small but thriving manufacturing base dominated by craft forms of production serving the local area. Despite a history of vigorous workers' struggles, Los Angeles had also become a pre-eminent centre of effective labour control, an area where the open shop was virtually a law in the fifty years following the 1890s depression.[3] The persistent strength of anti-union employers groups and their supportive public sector allies has been central to the twentieth-century industrialization of Los Angeles.

By the 1920s, Los Angeles was already the most automobile-oriented city in the world, an inclination which both reflected and contributed to its highly decentralized urban morphology. Not surprisingly, local craft manufacturing focused early on innovative and experimental cars, modish vehicle equipment, and specialized car design (Morales, 1986). At about the same time, there also developed a significant aircraft industry, fostered in part by a particularly suitable physical environment but, even more, by its association with small-scale specialized car and machinery production. This local craft tradition, significantly spurred by the relative isolation of Los Angeles from the major industrial centres and markets of the north-east, laid the foundation for the Fordist urban industrialization of the interwar years (as it would elsewhere in the USA).

During the depression, four major car manufacturers opened mass production assembly plants in Los Angeles. By the late 1940s, Los Angeles had developed the largest mass market automobile–glass– rubber tyre manufacturing complex outside the Midwest. This propitious combination of craftsmanship and mass production, automobile and aircraft manufacture, abundant supplies of labour and petroleum, increasing federal expenditures for defence and suburbanization, and the beneficence of sunshine and the open shop sustained rapid economic expansion through the depression years and stimulated extraordinary war-based booms, first in the Second World War and then during the Korean War. From 1950–53, the Korean War years, total employment increased by 415,000 jobs, 95,000 in the aircraft industry (by then

---

3. Perry and Perry (1963, vii) claim that 'With the possible exception of San Francisco in the 1920s, it is doubtful if the labour movement has ever faced anti-union employer groups so powerful and well organized as those in Los Angeles.'

reoriented from an emphasis on aircraft frames to a more diversified aerospace–electronics–guided missile manufacturing).

The accelerated industrialization of Los Angeles during the middle third of the century was partly an extension of the Fordist urban–industrial development that then characterized the American Manufacturing Belt. Craft production continued to be highly significant, but in many industrial sectors Fordist labour processes and the mass production of consumer durables grew as rapidly as they did anywhere else in the country. Los Angeles came to epitomize the state-managed industrial metropolis. It was an exemplary arena not only for Fordist industrial expansion but also for Keynesian demand stimulation and mass consumerism; the pump-priming federal management of macro-economic variables; the labour-stabilizing creation of productivity-linked social contracts between government, business, and unions; and, perhaps most paradigmatically, the sprawling suburbanization that pushed urban life into the automaniacal and generously state-subsidized 'crabgrass frontier' (Jackson, 1985).

What was particularly distinctive about Los Angeles during this period was that it simultaneously developed as a prototype for what is now commonly described as 'sunbelt' forms of industrialization and urban growth. In particular, its burgeoning aerospace industry grew to become the focus of a high-technology industrial complex combining civilian aircraft manufacturing with advanced electronics, space exploration, weapons research, and massive national defence contracting. By the 1960s, when the post-war boom was reaching its last crescendo, the largest technopolis in the country had become firmly established in the region. There thus came to be juxtaposed within Los Angeles highly advanced forms of both the older Fordist and newer postfordist industrialization processes and regimes of accumulation, each closely tied to federal programmes and expenditures.

This double-barrelled industrialization became localized in a low density sprawl of residences and workplaces linked by a network of freeways and set into an extraordinary fragmentation of political jurisdictions, literally hundreds of local governments (Miller, 1981). The downtown core of the city of Los Angeles remained the largest node, but its relative size advantage diminished significantly with the accelerated growth of regional shopping centres and suburban industrial sites. In response to this decline, powerful corporate interests flexed their muscles again, crushing what had promised to be one of the largest public housing programmes in the country under the banner of 'fighting socialism' (Parson, 1982). The 'public interest' and public expenditures were shifted to major renewal programmes aimed at reviving the central city business district and selectively gentrifying the extensive areas of deteriorated housing surrounding it, again reflecting (if not setting) national urban trends.

One of the largest urban industrial zones in the world still stretched southward from downtown, cutting through rigidly segregated areas of Black and poor White workers, tens of thousands of whom had migrated there during the 1930s and 1940s. But here too, in this most comfort-able-looking of blue-collar suburbias, there was relative decline as new industrial and residential expansion spilled outward to Orange and other surrounding counties, leaving behind one of the starkest racial employ-ment divides anywhere in the country. West of the Alameda 'White Curtain' was the compacted Black ghetto, the third largest in the coun-try, tantalizingly close to but increasingly distanced from the large pool of jobs to the east.

As perhaps the quintessential centre of state-managed and locally boostered capitalist urbanization, Los Angeles also epitomized the crises that this urbanization helped to generate. The Watts riots in 1965 explosively initiated a series of urban challenges to the legitimacy of the post-war economic order in the USA and Western Europe. Los Angeles' large Chicano population also responded; first through high school boycotts in 1966 and then, in 1970, in the Chicano Moratorium, an anti-war protest of 30,000 people, probably the largest Latino poli-tical demonstration in recent American history. The fiscal strains and social tensions of the state-managed metropolis were clearly showing, especially in Los Angeles County. Total welfare payments more than doubled between 1964–69 while aid to families with dependent children trebled. The shock of the global recession of 1973–75 merely confirmed what had already been recognized in Los Angeles and elsewhere: that the continuation of post-war economic growth could no longer depend on business as usual. A far-reaching and fixative reorganization of the social and spatial structures of accumulation was necessary. Another era of intensified competition between the old and the new, between an inherited and projected order, between continuity and change had begun.

## The Spatial Restructuring of Los Angeles

The changing sectoral distribution of employment offers revealing indi-cations of the scope and intensity of spatial restructuring in the Los Angeles urban region over the past twenty years. As can be seen from Figure 2, the services sector has shown the highest rate of employment increase in both the 1960s and the 1970s, and from all indications probably continues to lead all other sectors in the 1980s. Although its rate of growth has slackened since 1970, total employment in the services sector recently surpassed employment in manufacturing to become the largest employment sector in the region, a position it last held in the 1920s.

This rapid growth of jobs in services as well as in wholesale and retail trade and FIRE (finance, insurance and real estate), along with the marked relative decline in government employment (especially federal),

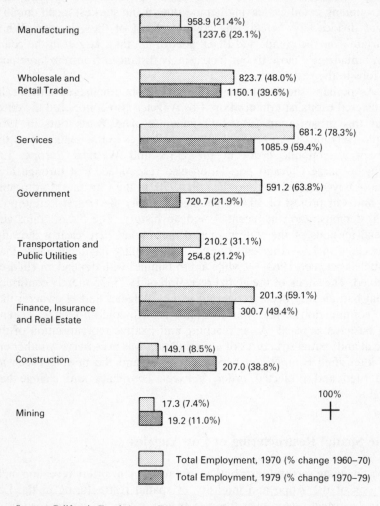

Manufacturing — 958.9 (21.4%) / 1237.6 (29.1%)

Wholesale and Retail Trade — 823.7 (48.0%) / 1150.1 (39.6%)

Services — 681.2 (78.3%) / 1085.9 (59.4%)

Government — 591.2 (63.8%) / 720.7 (21.9%)

Transportation and Public Utilities — 210.2 (31.1%) / 254.8 (21.2%)

Finance, Insurance and Real Estate — 201.3 (59.1%) / 300.7 (49.4%)

Construction — 149.1 (8.5%) / 207.0 (38.8%)

Mining — 17.3 (7.4%) / 19.2 (11.0%)

100%

Total Employment, 1970 (% change 1960–70)
Total Employment, 1979 (% change 1970–79)

Source: California Employment Development Department

*Figure 2*  Sectoral employment changes 1960–1979

parallels broad national trends. Where Los Angeles most markedly departs from these national trends is in the pattern of change between the two decades, in particular the continuing expansiveness of the manufacturing sector. To appreciate the relative intensity of these shifts since 1970 and their geographical variation within the region, it is useful to refer to Figure 3, which presents the results of a shift-share analysis of employment changes over three time periods, 1962–67 (before any significant restructuring had taken place), 1967–72 (a transitional period), and 1972–77 (after a first round of change).[4]

REGIONAL SHARE INDEX, 1962–77*

|  | Region | L.A. County | Orange | S.B./Riv. |
|---|---|---|---|---|
| Manufacturing | 92033 <br> + − + | −25803 <br> − − + | 96310 <br> + + + | 21515 <br> + + + |
| Wholesale and Retail Trade | 41429 <br> + + + | −81157 <br> − − − | 99639 <br> + + + | 22946 <br> + + + |
| Services | −20544 <br> − − + | −113500 <br> − − − | 79863 <br> + + + | 13092 <br> + + + |
| Government | N.A. | N.A. | N.A. | N.A. |
| Transportation and Public Utilities | 9411 <br> + + − | −10647 <br> + − − | 13663 <br> + + + | 6395 <br> + + + |
| Finance, Insurance and Real Estate | 33692 <br> + + + | −5250 <br> + + − | 33741 <br> + + + | 5201 <br> + + + |
| Construction | −34980 <br> − + + | −47348 <br> − − − | 12616 <br> − + + | −248 <br> − + + |
| Mining | −4439 <br> + − − | −4788 <br> + + − | 104 <br> + − + | 244 <br> + − − |

*Shift Share Analysis covers only 4 counties, omitting Ventura. + and − below index indicate positive or negative growth over 3 time periods (1962–67, 1967–1972, and 1972–77)

*Figure 3*   Shift-share analysis of employment change 1962–1977

---

4. See Morales, et al. (1982). Shift-share analysis breaks down the actual employment change in a sector into three components: national growth (the number of jobs that would have been added had the sector grown at the aggregate national growth rate across all sectors); industrial mix (the number of jobs that would have been added or lost assuming that the sector changed at the same rate as it did at the national level); and regional share (the derived or particularly local change in employment after national trends are accounted for). Using manufacturing employment 1962–77 as an illustration, the three indices are 393,853 (reflecting the national employment growth rate of 48 per cent); −265,598 (showing the negative growth of manufacturing nationwide); and 92,033 (an indication of the degree to which the region departed from national norms). Figure 3 contains only the particularizing regional share indices.

For the region as a whole, trade and FIRE have positive regional share indices over the three periods, indicating that regional employment grew faster than national averages. Growth in services and construction, however, was slower, resulting in a composite negative regional share (although both pick up strength in 1972–77). The largest overall index by far is for manufacturing, based primarily on the explosive growth of the Orange County industrial complex (Scott, 1986). Los Angeles County has a negative share over the whole time period, but manufacturing jobs are positive in the key 1972–77 years, suggesting a relative recovery of manufacturing growth. These measurements depict an internally differentiated and increasingly decentralized regional employment structure with an unusually robust manufacturing sector. A tertiarization of the regional economy is proceeding rapidly, but primarily via wholesale and retail trade and finance, insurance, and real estate rather than the services sector. The peripheral counties, especially Orange, seem to be booming in employment growth in virtually every sector, whilst the core county of Los Angeles declines rapidly at first but appears to begin a minor recovery towards the end of the period.

What lies behind this localized sectoral restructuring and how has it affected the spatialization of the region? A series of thematic arguments can be synthesized from the flood of recent research on the changing political economy of Los Angeles and used to describe the underlying dynamics of the contemporary urban and regional restructuring. The arguments are rooted in what appear at first to be paradoxical juxtapositions, combinations of contraries, which with more detailed elaboration become more understandably combinable and articulated rather than rigidly oppositional. The organization of the discussion thus suggests an interpretive framework which can be used to examine the impact of urban restructuring in other major metropolitan regions, indicatively viewed through the window provided by the specific recent experience of the Los Angeles space economy and its postfordist transformation.

### Deindustrialization and reindustrialization

Frostbelt and Sunbelt dynamics come together in Los Angeles, intermeshing to produce a complex mix of selective industrial decline and rapid industrial expansion. This has seemingly condensed the separate recent experiences of Detroit and Pittsburgh as well as Houston and the Silicon Valley to create a strikingly different industrial geography than that which had developed before 1960. Earlier rounds of industrialization had concentrated production and employment within a broad zone stretching south from the city centre of Los Angeles to the twin ports of San Pedro and Long Beach, with important outliers in the San

Fernando and San Gabriel Valleys and in the so-called Inland Empire of San Bernardino County (once the site of the largest steel complex west of the Mississippi River). Within this extensive urban industrial landscape, whole municipalities, such as Vernon and the bluntly named City of Industry and City of Commerce, became entirely devoted to manufacturing and related warehousing and commercial services, with almost no local population to get in the way. Almost 50,000 people worked in Vernon, for example, but less than one hundred lived there. Other cities such as South Gate (next door to Watts and almost midway between downtown and the ports) mixed heavy industrial production with some of the most attractive working-class residential neighbourhoods in the country. These were almost entirely white neighbourhoods, it should be added, for running through this industrial zone was one of the most rigidly defined racial divides in any American city.

Today, these areas have become the rustbelt of Los Angeles, with numerous abandoned factories, high unemployment rates, economically devastated neighbourhoods, extensive outmigration, and deskilling and wage-reducing occupational shifts from industry to service jobs. What had once been the second largest automobile assembly complex in the country has been reduced to a single General Motors plant in Van Nuys, currently the site of a vigorous labour and community struggle against plant closure (Mann, 1987). The second largest tyre manufacturing industry in the country has virtually disappeared, with Firestone, Goodyear, Goodrich, Uniroyal and other smaller firms closing down shop, accompanied by much of the Southern California steel industry. In the four years 1978–82, at least 75,000 jobs were lost due to plant closings and indefinite lay-offs, affecting primarily a segment of the labour market that was highly unionized and contained an unusually large proportion of well-paid minority and female blue-collar workers. Figure 4 maps this pronounced deindustrialization.

In South Gate since 1980, the closure of Firestone Rubber, General Motors, and Norris Industries–Weiser Lock has meant the loss of what, at the peak employment of earlier years, was over 12,500 jobs. In adjacent Watts, economic conditions have deteriorated more rapidly than in any other community within the City of Los Angeles and are now no better – and probably worse – than they were at the time of the Watts riots in 1965. In the decade and a half following the insurrection, the predominantly Black area of South-Central Los Angeles lost 40,000 in population, the labour force was reduced by 20,000, and median family income fell to $5900 – $2500 below the city median for the Black population in the late 1970s.

This selective deindustrialization has had a major effect on the overall strength of organized labour. During the 1970s, unionization rates in

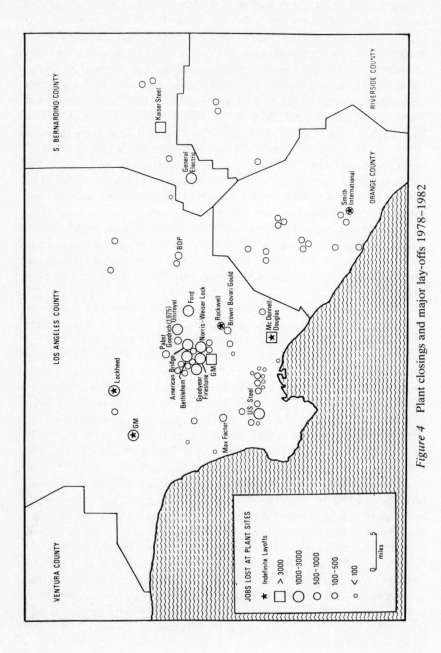

*Figure 4* Plant closings and major lay-offs 1978–1982

Los Angeles County dropped from over 30 per cent to about 23 per cent. In Orange County, the drop was even more pronounced and, in manufacturing, which was experiencing an extraordinary expansion of employment, unionization rates plummeted from 26.4 to 10.5 per cent. This represented an absolute decline of over one quarter of the union membership in 1971. Thus, both deindustrialization and new industrial growth became associated with significant declines in unionization and a weakening of many of the contractual gains achieved by organized labour in the two decades following the Second World War.

Selective deindustrialization and wholesale deunionization in Los Angeles extended national and global trends generated in response to the economic and political crises which shattered the post-war boom and the Keynesian social contracts that were so much part of the expansionary period. As occurred during similar periods of crisis and restructuring in the past, technological innovations, corporate-managerial strategies, and state policies became more directly and explicitly focused on two increasingly vital and closely related objectives: the restoration of expanding superprofit opportunities and the establishment of more effective control over the labour force. Under the rationale of an urgently needed 'rationalization', a renewed celebration of the creative destruction crucial to the survival of capitalist economies, the economic landscape that consolidated during the post-war years began to be selectively destroyed and equally selectively reconstructed.

This comprehensive attempt to discipline labour (along with less efficient capitals and major segments of the central and local state) has been an essential motif in the contemporary restructuring process. It is being promulgated by a varied range of options revolving primarily around the relative reduction of labour costs. These options include direct attacks on the most powerful working-class organizations, the increasing centralization and concentration of capital, intensified capital mobility to establish a constant threat of closure and relocation, induced technological innovation to cut costs and create improved instruments for labour control and productivity, growing subsidization of large corporations by the federal state and by local governments devoted to attracting new employment, all wrapped up in ideological programmes designed to justify sacrifice and austerity by some (ostensibly for the greater good) while force-feeding others in the valiant hope that the excess will somehow trickle down to the awaiting underclasses. These are the primary pathways of restructuring, whether we refer to the economy as a whole or to an individual corporation. The strategies are not always successful, but the driving objectives cannot be overlooked amidst all else that is changing in the contemporary world.

This more focused interpretive perspective on restructuring makes the deindustrialization of Los Angeles much more than the local expression of an innocent process of modernization and 'post-industrial' evolution, a minor accompaniment to an otherwise booming local economy. Deindustrialization has been a critical fulcrum around which many other aspects of social and spatial restructuring revolve (Bluestone and Harrison, 1982). It forms the necessary backdrop, for example, to understanding the remarkable industrial expansion that has been taking place in the Los Angeles region since the 1960s. Let us examine more closely the two most propulsive contexts for this expansive industrial growth: aerospace/electronics and apparel manufacturing.

Figures 5 and 6 give a picture of the magnitude and locational pattern of what may be the largest urban concentration of advanced technology-based industry in the world. During 1972–79, employment in the aerospace/electronics cluster of seven industrial sectors grew by 50 per cent in the region, with an addition of over 110,000 jobs and a rise in its percentage of total manufacturing employment from 23 per cent to 26 per cent. Employment growth in these seven sectors was greater than the total increment of new manufacturing jobs in Houston during the same period and represents almost the equivalent of the net addition of the entire high technology labour force of the Silicon Valley, 110,000 to the 147,000 jobs in the same seven sectors in Santa Clara County (1979). More recent data show that by 1985 Los Angeles County alone employed over 250,000 workers in the Bureau of Labor Statistics 'Group 3' category, another widely used definition of high technology industry. The equivalent figure for Santa Clara County was 160,000.

What ties together the sectors which comprise this technopolitan industrial complex more than anything else has been their shared dependency upon technology arising from Department of Defense and NASA research and the stimulus of military contracts. Southern California has been by far the largest recipient of prime defence contracts since the 1940s. The continued expansion of high technology industries in the 1980s has been promoted still further by the military-centred Keynesianism of the Reagan administration, with Los Angeles County in particular being one of the leading beneficiaries of research funding for Star Wars, Reagan's Strategic Defense Initiative.

This high technology expansion in Los Angeles bears some comparison with what has been occurring in the region around Boston, where a deep and prolonged process of labour disciplining, plant closures, and capital flight produced the conditions for an economic recovery based in large part on rapidly expanding high technology and service sectors (Harrison, 1984). In both urban regions, an occupational recycling has been taking place, increasingly polarizing the labour market by skill and

| SIC Code | Sector | | Total Employment: 1972 | Total Employment: 1979 | % US Employment in sector 1972 | % US Employment in sector 1979 |
|---|---|---|---|---|---|---|
| 372 | Aircraft and Parts | Region | 108,501 | 100,956 | 21.8 | 19.2 |
| | | LA Cty | 103,076 (95.0) | 90,153 (89.3) | | |
| | | Orange | 3,581 ( 3.3) | 7,369 ( 7.3) | | |
| | | SB/R/V | 1,844 ( 1.7) | 3,434 ( 3.4) | | |
| 376 | Guided Missiles and Space Vehicles | Region | not a separate SIC category | 56,805 | — | 44.4 |
| | | LA Cty | | 47,297 (83.3) | | |
| | | Orange | | 7,500 (13.2) | | |
| | | SB/R/V | | 2,008 ( 3.5) | | |
| 357 | Office and Computing Machines | Region | 20,146 | 30,967 | 9.2 | 9.1 |
| | | LA Cty | 15,815 (78.5) | 14,431 (46.6) | | |
| | | Orange | 3,969 (19.7) | 15,886 (51.3) | | |
| | | SB/R/V | 362 ( 1.8) | 650 ( 3.5) | | |
| 365 | Radio and TV Equipment | Region | 8,016 | 8,695 | 6.6 | 9.3 |
| | | LA Cty | 8,016 (100) | 7,514 (86.4) | | |
| | | Orange | — | 1,181 (13.6) | | |
| | | SB/R/V | — | — | | |
| 366 | Communications Equipment | Region | 50,179 | 64,158 | 11.7 | 12.1 |
| | | LA Cty | 27,699 (55.2) | 36,698 (57.2) | | |
| | | Orange | 22,480 (44.8) | 25,984 (40.5) | | |
| | | SB/R/V | — | 1,476 ( 2.3) | | |
| 367 | Electronic Components and Accessories | Region | 28,043 | 53,384 | 8.6 | 11.4 |
| | | LA Cty | 19,715 (70.3) | 29,308 (54.9) | | |
| | | Orange | 4,879 (17.4) | 17,510 (32.8) | | |
| | | SB/R/V | 3,449 (12.3) | 6,566 (12.3) | | |
| 382 | Measuring and Controlling Devices | Region | 7,224 | 17,485 | 7.9 | 8.1 |
| | | LA Cty | 7,224 (100) | 12,807 (73.2) | | |
| | | Orange | — | 4,266 (24.4) | | |
| | | SB/R/V | — | 412 ( 2.4) | | |
| | TOTAL | Region | 222,109 | 332,450 | | |
| | | LA Cty | 181,545 (81.7) | 238,208 (71.6) | | |
| | | Orange | 34,909 (15.7) | 79,696 (24.0) | | |
| | | SB/R/V | 5,655 ( 2.6) | 14,546 ( 4.4) | | |
| | Total as % of Regional Manufacturing Employment | | 23 | 26 | | |

Source: Department of Commerce, *County Business Patterns*, 1972 and 1979.

*Figure 5* Employment change in the aerospace/electronics industries

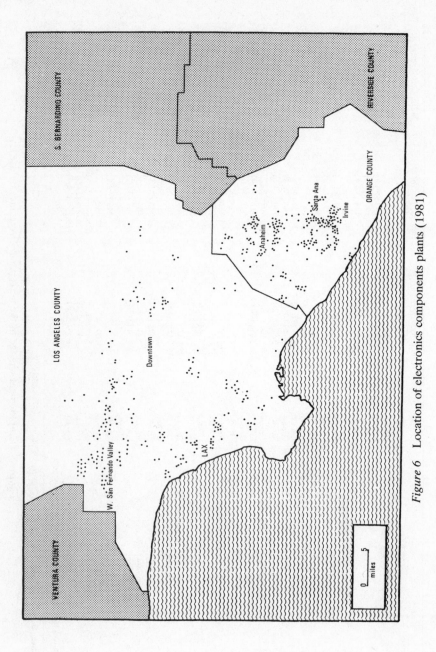

*Figure 6*  Location of electronics components plants (1981)

wage differentials. The middle segment of skilled, unionized, and well paid blue-collar workers has been shrinking, with a small number of its expelled labourers floating up to an expanded white-collar technocracy but a much larger proportion percolating downward into a relatively lower-skilled and lower-wage reservoir of production and service workers, swollen by massive immigration and part-time and female employees. This downward percolation of formerly well established union workers was once described as a demoralizing 'K-Marting' of the labour market.[5]

The growth of the garment industry reflects another dramatic change in the regional labour market. Not only has the 'high technocracy' settled in extraordinary numbers in Los Angeles, but so too has what is probably the largest pool of low-wage, weakly organized, easily disciplined immigrant labour in the country. This still-growing labour pool has affected virtually every sector of the regional economy, including the aerospace and electronics industry. Its imprint has been most visible, however, in the production of garments, especially in the category of 'women's, misses' and juniors' outwear', which tends to be highly labour-intensive, difficult to mechanize, and organized around small shops to adapt more efficiently to rapidly changing fashion trends.

Employment in apparel manufacturing expanded by nearly 60 per cent between 1970–80, representing 12 per cent of total manufacturing employment growth and a net addition of over 32,000 jobs. Of the approximately 125,000 jobs in this industrial sector, perhaps as many as 80 per cent have been held by undocumented workers in recent years, with 90 per cent of all employees being women. Unionization rates are low and infringements of minimum wage, overtime, child labour and occupational safety laws are endemic. Sweatshops which provoke

---

5. When the President of the United Electrical Workers Local was asked a few years ago what jobs her 1000 members – the majority of whom were skilled women earning $10 to $12 an hour – would be able to find after General Electric closed down its flatiron plant in Ontario (San Bernardino County), she replied: 'Clerks at the local K-Mart store'. The struggle to prevent the closure of this General Electric plant, which had produced nearly all the metal steam irons sold by General Electric in the USA, was both vigorous and unsuccessful. It involved the formation of a coalition of labour, community, and religious leaders and the organizational and technical assistance of the Los Angeles Coalition Against Plant Shutdowns (LACAPS) to pressure state and local government to create or help increase the level of special services and programmes for displaced workers. Requests from the United Electrical Workers Local and from LACAPS to some members of the faculty of the Urban Planning Program at UCLA to help them make practical sense of the industrial changes then taking place in the Los Angeles region stimulated the first round of research on urban and industrial restructuring (including Soja, Morales and Wolff, 1983). The first product of this work – a pamphlet on 'Early Warning Signs of Plant Closure' designed to be distributed on the shopfloor of the General Electric plant – appeared on the day that General Electric formally announced the closure of its Ontario factory.

images of nineteenth-century London have thus become as much a part of the restructured landscape of Los Angeles as the abandoned factory site and the new printed circuit plant. And they can be found not only in the garment industry but in many other manufacturing sectors as well (Morales, 1983).

Both the deindustrialization and reindustrialization of Los Angeles are to some extent continuations of trends which began before the mid-1960s. Their acceleration, however, and their fortuitous interlinking and connection to other changes taking place in the region, have reshaped the sectoral structure of the regional economy over the past twenty years. Viewed as a three-tiered structure, the industrial labour market has been significantly squeezed in the middle layers, expanded somewhat at the top, and massively broadened at the bottom. To consider only this three-part division is not enough, however, for the segment-ation and recomposition of the labour market has been much more finely grained and complex. Cutting across the broad polarization, the occupational recycling, and the expansive new job generation has been a further fragmentation based on race, ethnicity, immigrant status, and gender. The end result today is a regional labour market that is more occupationally differentiated and socially segmented than ever before.

## Geographical decentralization and recentralization

Sectoral restructuring and increasing labour-market segmentation have been paralleled by an equally pronounced spatial restructuring and changes in the occupational and residential geography of the region. This reconfiguration of production and consumption patterns also appears at first to be paradoxical, for it has involved both a continuation of past trends of disagglomeration and a partial reversal of these trends, leading to the formation of new or renewed urban concentrations within the metropolitan space economy. As was the case with industrial restructuring, the integral logic of this combination of contraries becomes clearer when the aggregate picture is decomposed into its more intricate internal dynamics.

A sprawling and polynucleated decentralization process has charac-terized the historical geography of the capitalist city since the nineteenth century. In many ways, Los Angeles has been and continues to be an exemplary case of this decentralized urban/suburban growth. As the older industrial and residential subregions have declined, the regional periphery has expanded over the past twenty years at a rate that may be unsurpassed anywhere else in the country. The four counties of Orange, San Bernardino, Riverside, and Ventura collectively averaged a 40 per cent increase in population during the 1970s and experienced an even higher rate of employment growth. All available evidence suggests that these growth rates have been maintained in the 1980s. Significant expansion has also been occurring in the outer areas of Los Angeles County, accounting for a large part of its population increase of 450,000 in the 1970s and over 600,000 between 1980–85 alone.

Metropolitan decentralization today, however, has expanded well beyond the local region in Los Angeles, as it has in most other large cities in advanced industrial countries. More than ever before in recent American history, both population and industry have been moving into smaller towns and rural areas, evoking what some have called the 'great non-metropolitan turnaround', another of the characteristic features that have been ascribed to the contemporary restructuring process. The decentralization of urban industrial production, however, has progressed still further afield in the past two decades to take on an increasingly global expression, with relocating industrial firms skipping through the national hierarchy of urban and rural centres entirely to settle overseas in newly industrializing regions. This expanded turnaround has dramatically internationalized the scope of urban analysis and widened the context for understanding the implications of urban restructuring in the capitalist city (Smith and Feagin, 1987).

The rapid aggregate growth in jobs and employment in Los Angeles has masked somewhat a significant outmigration of population and industries. This, too, must be seen as part of the decentralization pattern, whether it be counted as the number of industrial establishments relocating or setting up branch plants outside the region, or as the growing volume of return migrants – Blacks moving back to the South (perhaps as many as 25,000 between 1975–80), undocumented workers moving back over the border to Mexico (possibly hundreds of thousands in the past two decades, although many return to Los Angeles again and again), and many others who leave having been unable to find affordable housing. The accumulated volume of outmigration from Los Angeles has been substantial, but it has been overbalanced by extraordinary levels of population and industrial growth.

Also disguised by the aggregate data on regional decentralization, and supplying another perplexing challenge to conventional urban analysis, has been an unprecedented recentralization of economic activities in Los Angeles. This has taken two main forms, each reflecting both the internationalization of the urban space economy and the composite of technological and organizational changes that have marked the rise of flexible specialization in the production of goods and services (Storper and Christopherson, 1987). The two forms can be popularly characterized as 'downtown renaissance' and the rise of the 'outer city'.

After decades of public and private campaigns to defy the claims that

Los Angeles was little more than a hundred suburbs in search of a city, there has developed in the past twenty years a visible and expansive downtown core to the giant regional metropolis. And if one takes this downtown core to include a twenty-mile extension along the Wilshire Boulevard corridor to the Pacific, it is today a central city that is almost commensurate with the size and scope of the regional economy. By the 1960s, downtown Los Angeles was already the site of the second largest concentration of government employment in the country and was an important domestic banking and financial centre (although clearly behind San Francisco). Since then, it has expanded dramatically as a governmental and corporate citadel, a commercial and industrial nucleus, and a control centre for both domestic and international capital.

Again, Los Angeles provides an exaggerated case of more general national trends, an extension of the renaissance of once deteriorating downtowns and the office-building booms that have constructed new skylines in the centres of sunbelt (and some frostbelt) cities. Built into the new centrality of downtown Los Angeles has been a primary geographical locus for the accelerated centralization, concentration, and internationalization of industrial and finance capital that has marked the contemporary restructuring of the world economy. Positioned increasingly as a 'capital of capital' in the Pacific Basin, Los Angeles has been surging toward the ranks of the three other capitals of global capital, New York, London, and Tokyo (its Pacific Rim cohort).

Figure 7 presents the corporate-financial citadel of central Los Angeles. In the zone stretching from downtown to the ocean and branching south to the airport (LAX) are over sixty major corporate headquarters, a dozen banks and savings and loan companies with assets over one billion dollars, five of the eight largest international accounting firms, two-thirds of the 200 million square feet of high-rise office space in the region, a battalion of corporate law offices unrivalled off the east coast, and the national nucleus of the American military-industrial complex. Also contained within the same area are condominiums which are advertised for $11 million (with a Rolls-Royce thrown in for free) and the largest concentration of homeless people in the country, a reminder of the intense economic polarization associated with the restructuring of Los Angeles. In many of the residential areas adjacent to the extended downtown, population densities now approximate those in the large cities 'back east'.

Another dramatic and polarizing recentralization has been taking place in Orange County, where at least 1,500 high technology firms have clustered since the mid-1960s. An outer city of paradigmatic proportions, the Orange County complex has become an important

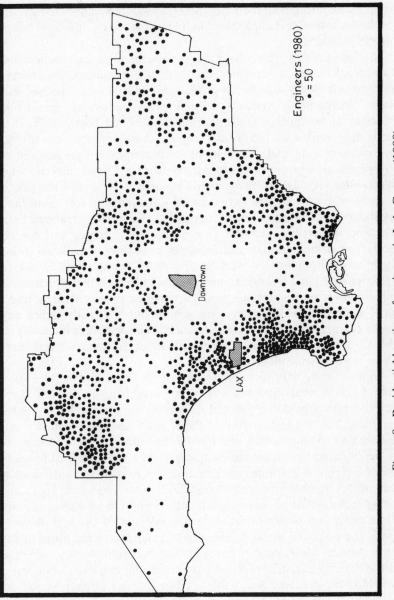

*Figure 8* Residential location of engineers in L.A. County (1980)

laboratory (along with Silicon Valley and Route 128 in Boston) for investigating the internal dynamics of the new urban growth centres taking shape throughout the USA (Scott, 1986). In Orange County, the postfordist urban landscape takes on its most representative and symptomatic contours.

The formative impulse behind this 'peripheral' urbanization, here and elsewhere, has been the creation of a dense nest of transactional linkages and technologically advanced production and service systems that enable increasingly vertically disintegrated industrial production processes to be flexibly and efficiently re-attached horizontally, in a burgeoning territorial industrial complex. Amalgamated around this agglomerative industrial system are the accoutrements of the new silicon landscapes: high-income and expensively-packaged residential developments; huge regional shopping centres reputed to be among the largest in the world; created and programmed environments for leisure and entertainment (epitomized by Disneyland in Anaheim); organized links to major universities and the Department of Defense; and several enclaves of cheap and manipulable labour constantly replenished by in-migration of both foreign workers and those deindustrialized out of higher paying jobs. Orange County has been one of the prefigurative technopolises, an amorphous regional complex that confounds traditional definitions of both city and suburb but insistently defies any simple characterization as 'post-industrial'. The centre-less outer city is, after all, as much the product of industrialization as the urban agglomerations of the nineteenth century.

Another exemplary technopolis has taken shape in the sub-region around Los Angeles International Airport, reaching south from Santa Monica to the armed and guarded residential villas of the Palos Verdes peninsula. This is the heartland of the region's aerospace industry and defence contracting, as well as a rapidly expanding centre for electronics, major banks and insurance companies, and a wide range of business services. Figure 8 illustrates the extraordinary residential concentration of engineers that has grown up around LAX to serve the local high technology industries. It peaks just south of the airport in an almost entirely White racial and occupational enclave, a redoubt of the high technocracy. The map also shows a clustering of engineers to the north of the Santa Monica Mountains in the west San Fernando Valley, where a third outer-city industrial complex has been taking shape in recent years; and along the south-western border of Los Angeles County, in part an extension of the Orange County labour market nearby.

The partitioned occupational geography of Los Angeles County is broadly summarized in Figure 9. Superimposed over a relatively evenly distributed white-collar population is a pronounced residential

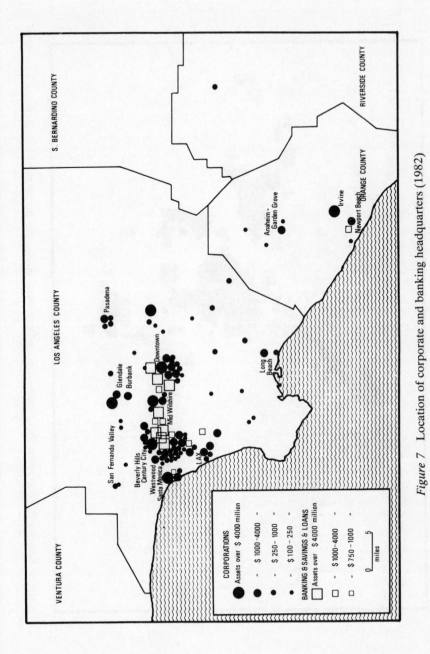

*Figure 7* Location of corporate and banking headquarters (1982)

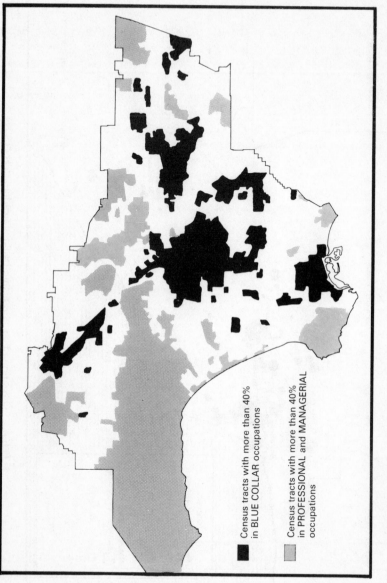

■ Census tracts with more than 40% in BLUE COLLAR occupations

░ Census tracts with more than 40% in PROFESSIONAL and MANAGERIAL occupations

*Figure 9*  Residential concentrations of blue-collar and executive/managerial occupations in L.A. County (1980)

polarization between the blue-collar working class in the older central cities and the peripheral retreats of the executive, managerial, and supervisory controllers of labour power on the hillsides and beaches. And within each of these concentrically opposing blocs, there is further residential specialization and enclosure, keeping everyone in their place.

The combined force of geographical decentralization and recentralization has contributed significantly to the increasing labour discipline that has accompanied the spatial restructuring of the Los Angeles region. Greater sectoral segmentation in the labour market has been accommodated and reinforced by a geographical fragmentation and segregation of the labour force at both the place of work and the place of residence. This changing urban geography recalls Poulantzas's description (1978, 107) of the production and reproduction of capitalist spatiality:

> Social atomization and splintering ... a cross-ruled, segmented and cellular space in which each fragment (individual) has its place ... separation and division in order to unify ... atomization in order to encompass; segmentation in order to totalize; closure in order to homogenize; and individualization in order to obliterate differences and otherness.

## The peripheralization of the core

As has already been noted, the restructuring of the Los Angeles urban region has been closely associated with an increasing internationalization of the local economy. Foreign capital has been migrating into Los Angeles at an ever-expanding rate, buying land, building office complexes, investing in industry, hotels, retail shops, restaurants, and entertainment facilities. More than half the prime properties in downtown Los Angeles is now owned by foreign corporations or by partnerships with foreign companies, led by Japan and Canada; and foreign capital is said to have financed as much as 90 per cent of recent multistorey building construction. Perhaps only in New York City has there been such a massive urban shopping spree by international capital in so short a time and from so many different sources.

Carried along with this international outreach and inflow of capital investment and interest, and the emergence of Los Angeles as a major financial centre for global capitalism, has been a pronounced peripheralization of the regional labour market. Over the past twenty years, about two million people from Third World countries have moved into Los Angeles and extended their influence into virtually every aspect of the changing urban landscape and culture. The centre has been increasing its centred-ness, but it is simultaneously becoming peripheralized as well. Here is another apparent paradox of restructuring demanding to be explained.

Thirty years ago, Los Angeles County was 85 per cent non-Hispanic White, or 'Anglo'. Even then, however, there existed within the county one of the most rigidly defined Black ghettoes in the country and the largest urban concentration of Mexicans outside their homeland. Moreover, this minority population had already been instrumental in earlier rounds of industrial development as a source of exploitable cheap labour and as an effective tool for controlling more organized union workers. During the Great Depression, for example, their presence, in conjunction with the powerful coalition of anti-union interests, helped to squeeze the more established and expanding workforce both from 'below' (with respect to the vertical structure of the labour market) and from 'next door' (via the adjacency of the Black ghetto and the Latino *barrio* to the major zones of industrial growth). The relative quiescence of organized labour in Los Angeles during the waning years of the Depression is at least partially explained by these effective manipulations, backed by the maintenance of the open shop and an extensive system of labour contracting.

An intense episode of McCarthyist 'red-baiting' in the 1950s was indicative of the persistent strength of the labour movement. Despite its pre-war setbacks, organized labour in Los Angeles appeared to be on the verge of receiving one of the largest public housing programmes in the USA. By the early 1960s, however, these ambitious programmes had been re-routed into urban renewal and the strength of the anti-labour 'growth machine' had become even more firmly established. In Los Angeles and elsewhere, these early post-war events signalled the progressive weakening of the New Deal connection between party politics and organized labour, and the concurrent rise of movement politics, mobilized locally in response to the destructive effects of urban redevelopment (Parson, 1982, 1985). The most shattering challenge to the local regime of accumulation that had sustained fifty years of rapidly expanding industrial production in Los Angeles thus came not from organized labour as much as from the formerly 'instrumental' minority populations. If there is just one event to choose from to signal the initiation of the contemporary restructuring process all over the world, it may very well be the Watts riot of 1965.

Viewed this way, the contemporary restructuring of the Los Angeles region has many of the characteristics of a restoration, an attempted reestablishment and refurbishing of a system of labour relations which had proved successful in the past. In this round of restructuring, however, the reserve army of minority and migrant workers (augmented by a massive entry of women into the workforce) has grown to unprecedented levels, creating an overflowing pool of cheap, relatively docile labour that is not only locally competitive but also able to compete with the new industrial concentrations of the Third World.

The magnitude and diversity of immigration to Los Angeles since 1960 is comparable only to the New York-bound wave of migrants at the turn of the last century. A resurgence of migration from Mexico has added at least a million residents to the existing population and helped to shade into the regional map a nearly ubiquitous Spanish-speaking presence. Since 1970, more than 200,000 Koreans have settled in Los Angeles and become a major influence in retail trade and the garment industry, with Korean family labour proving highly competitive with even the worst sweatshops. Filipinos, Thais, Vietnamese, Iranians, Guatemalans, Colombians and Cubans have arrived in large numbers and several Pacific Island populations have grown almost as numerous as in their home areas. In what may be the record for in-migration from a single country, nearly 400,000 Salvadorans are estimated to have moved to Los Angeles since 1980. Today, the Anglo population of the City and County of Los Angeles has become the minority. Figure 10 shows some of the major ethnic concentrations based on the 1980 census.

This demographic transformation has produced within the region many of the labour conditions and corporate advantages of Third World Export Processing Zones. Under certain market conditions and wage rates, especially when transport costs for similar goods produced overseas are sufficiently high, local shops in Los Angeles are producing automobile parts which are stamped 'Made in Brazil' and clothing marked 'Made in Hong Kong', enabling them to intervene in foreign delivery contracts. There is heavy competition in Los Angeles to establish new enterprise zones in high unemployment areas, but they will need unusually high rates of subsidization to attract new firms. The urban landscape is already filled with equivalent opportunities without the formal legislation.

The centre has thus also become the periphery, as the corporate citadel of multinational capital rests with consummate agility upon a broadening base of alien populations. The city that more than any other has been built upon the military defence of American shores has become the beach-head for a peripheral invasion. All that was local becomes increasingly globalized, all that is global becomes increasingly localized. What has been the political response to this extraordinary restructuring?

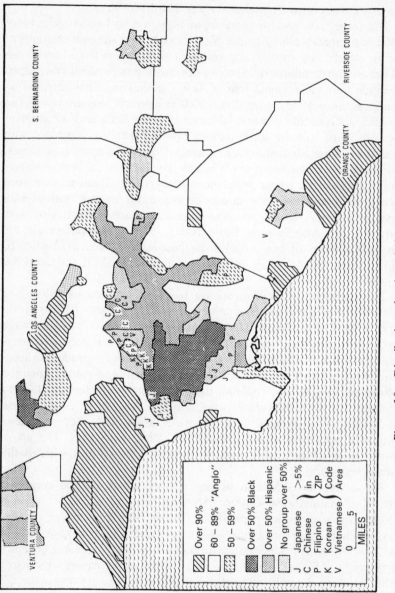

*Figure 10* Distribution of major ethnic groups (1980)

*Cacophonous silence: some observations on the politics of
restructuring*

For the most part, the restructuring of Los Angeles has proceeded with
remarkably little resistance. Organized labour in manufacturing has been
decimated, and earlier efforts to fight against threatened plant closures
have been virtually swallowed up in the great Los Angeles job machine.
Only a progressive faction of workers at the Van Nuys General Motors
plant continues the earlier momentum, struggling against continuing
lay-offs and the fragmenting effect of the introduction of 'flexible' team-
work labour contracts. Public employee labour unions remain fairly
strong and there are signs of increasingly successful labour organizing in
the garment industry and in various low-skill service sectors dominated
by immigrant workers. This continued strength is significant, but it pales
in comparison to the entrenchment of even more flexible and adaptive
coalitions of pro-growth interests blinded to the plight of labour by the
spectacular successes of the postfordist regional economy. The labour
movement, with a few exceptions, remains in a Fordist mode fighting
against an enemy that has become too slippery and diffuse to negotiate
with in traditional ways.

The ethnic politics that exploded in the 1960s and early 1970s has
been strangely silent ever since. The Black population has been increasingly polarized and fragmented into what are perhaps the richest and
poorest urban Black communities in the country, a polarization that has
weakened Black political influence, despite the presence of a long-term
Black mayor and continued city council representation. The Los
Angeles City Council today also has Latino and Asian members, breaking through the blockage that has existed throughout most of the
restructuring period, but it is still too early to know whether and how
this will affect urban politics.

Perhaps the most dramatic effect of the Third World demographic
expansion has been the growth of a massive underground or informal
economy, but by its very nature the participants are shielded from direct
political participation. Aimed at surviving – and perhaps even prosper-
ing from – restructuring rather than publicly attacking its polarizing and
impoverishing instrumentality, the underground economy thrives in the
interstices of urban life, succouring ethnicities and providing the necessary niches for personal survival. Only when it breaks out from its niches
into wider networks of criminal gang- and drug-related activities does it
threaten the restructured order. The police response to these 'trespasses'
has been characteristically prompt and severe, but the rate of violent
crime remains very high and more than 20,000 people fill to overflowing
the official urban prisons of Los Angeles.

The most successful urban social movements have been organized around housing issues, a very basic response to the severe housing crisis that has accompanied restructuring. For a fleeting moment, the City of Santa Monica teetered on the verge of 'municipal socialism' as a progressive city council was elected by an aroused majority of renters bent on imposing rent controls. The rent control movement spread from what the apartment owners called the 'People's Republic' of Santa Monica to introduce similar legislation in other cities, including the City of Los Angeles and more recently into the gay- and grey-dominated new municipality of West Hollywood. But little else has come from these initial victories other than ephemeral inspiration.

Homeowners, as might be expected, formed much more powerful 'growth control' organizations, making the Los Angeles region perhaps the busiest hive of slow growth movements in the country. As the region densified, more and more groups mushroomed to attack the traffic snarls and visual blight affecting their protectable 'urban villages'. These defensive territorialities were not new, but the extraordinary intensity of urban development in Los Angeles provoked so many that, flushed with some renewed power, a few have begun to turn to broader issues than turf protection and simple NIMBY responses ('Not In My Back Yard'). With the region showing signs of being buried in sewage and other hazardous wastes, and being occupied by an army of the homeless, a more liberal tilt has recently become evident in these middle-class community movements, with new coalitions being formed with neighbourhood groups in poor, predominantly minority areas of the inner city. Today, these coalitions, especially when they become involved in such economic issues such as the minimum wage, occupational health and safety conditions, and affordable housing, pose perhaps the greatest perceived threat to the political establishment since the onset of restructuring.

These observations, however, merely skim the political surface. Underneath is a sea of intensifying urban stress that cuts across class, race, and gender to exemplify the social costs of relatively 'successful' restructuring and the still barely visible instrumentality of capitalist spatialization. How this restructuring and spatialization will be seen and responded to in the future remains an open question.[6]

---

6. Two analyses of local social movements in California have recently been published, providing some fresh insight into some of the general issues and strategies involved outside the Los Angeles region. See Molotch and Logan (1987) and Plotkin (1987). For an excellent analysis of contemporary urban politics in Los Angeles, see Mike Davis (1987).

**Finding Other Spaces?**

A concatenation of paradoxes has been used to describe Los Angeles and to exemplify the contemporary restructuring of the capitalist city, its deconstruction and tentative reconstitution. Ignored for so long as aberrant, idiosyncratic, or bizarrely exceptional, Los Angeles, in another paradoxical twist, has, more than any other place, become the paradigmatic window through which to see the last half of the twentieth century. I do not mean to suggest that the experience of Los Angeles will be duplicated elsewhere. But just the reverse may indeed be true, that the particular experiences of urban development and change occurring elsewhere in the world are being duplicated in Los Angeles, the place where it all seems to 'come together'.

The informed regional description that has been presented thus far, however, depicts only some of the broad vistas visible from the vantage point of Los Angeles. The seen/scene is that of a new geography of modernization, an emerging postfordist urban landscape filled with more flexible systems of production, consumption, exploitation, spatialization, and social control than have hitherto marked the historical geography of capitalism. Having put together this interpreted regional landscape, I will attempt to take it apart again to see if there are other spaces to be explored, other vistas to be opened to view.

# 9

# Taking Los Angeles Apart: Towards a Postmodern Geography

'The Aleph?' I repeated.
'Yes, the only place on earth where all places are – seen from every angle, each standing clear, without any confusion or blending' (10–11)

... Then I saw the Aleph.... And here begins my despair as a writer. All language is a set of symbols whose use among its speakers assumes a shared past. How, then, can I translate into words the limitless Aleph, which my floundering mind can scarcely encompass? (12–13)

(Jorge Luis Borges, 'The Aleph')

Los Angeles, like Borges's Aleph, is exceedingly tough-to-track, peculiarly resistant to conventional description. It is difficult to grasp persuasively in a temporal narrative for it generates too many conflicting images, confounding historicization, always seeming to stretch laterally instead of unfolding sequentially. At the same time, its spatiality challenges orthodox analysis and interpretation, for it too seems limitless and constantly in motion, never still enough to encompass, too filled with 'other spaces' to be informatively described. Looking at Los Angeles from the inside, introspectively, one tends to see only fragments and immediacies, fixed sites of myopic understanding impulsively generalized to represent the whole. To the more far-sighted outsider, the visible aggregate of the whole of Los Angeles churns so confusingly that it induces little more than illusionary stereotypes or self-serving caricatures – if its reality is ever seen at all.

What is this place? Even knowing where to focus, to find a starting point, is not easy, for, perhaps more than any other place, Los Angeles is everywhere. It is global in the fullest sense of the word. Nowhere is this more evident than in its cultural projection and ideological reach, its almost ubiquitous screening of itself as a rectangular dream machine for the world. Los Angeles broadcasts its self-imagery so widely that probably more people have seen this place – or at least fragments of it – than any other on the planet. As a result, the seers of Los Angeles have become countless, even more so as the progressive globalization of its urban political economy flows along similar channels, making Los Angeles perhaps the epitomizing world-city, *une ville devenue monde.*

Everywhere seems also to be in Los Angeles. To it flows the bulk of the transpacific trade of the United States, a cargo which currently surpasses that of the smaller ocean to the east. Global currents of people, information and ideas accompany the trade. It was once dubbed Iowa's seaport, but today Los Angeles has become an entrepot to the world, a true pivot of the four quarters, a congeries of east and west, north and south. And from every quarter's teeming shores have poured a pool of cultures so diverse that contemporary Los Angeles represents the world in connected urban microcosms, reproducing *in situ* the customary colours and confrontations of a hundred different homelands. Extraordinary heterogeneity can be exemplified endlessly in this fulsome urban landscape. The only place on earth where all places are? Again I appeal to Borges and the Aleph for appropriate insight:

> Really, what I want to do is impossible, for any listing of an endless series is doomed to be infinitesimal. In that single gigantic instant I saw millions of acts both delightful and awful; not one of them amazed me more than the fact that all of them occupied the same point in space, without overlapping or transparency. What my eyes beheld was simultaneous, but what I shall now write down will be successive, because language is successive. Nonetheless, I will try to recollect what I can. (Ibid., 13)

I too will try to recollect what I can, knowing well that any totalizing description of the LA-leph is impossible. What follows then is a succession of fragmentary glimpses, a freed association of reflective and interpretive field notes which aim to construct a critical human geography of the Los Angeles urban region. My observations are necessarily and contingently incomplete and ambiguous, but the target I hope will remain clear: to appreciate the specificity and uniqueness of a particularly restless geographical landscape while simultaneously seeking to extract insights at higher levels of abstraction, to explore through Los Angeles glimmers of the fundamental spatiality of social life, the adhesive relations between society and space, history and geography, the splendidly idiographic and the enticingly generalizable features of a postmodern urban geography.

## A Round Around Los Angeles

I saw a small iridescent sphere of almost unbearable brilliance. At first I
thought it was revolving; then I realized that this movement was an illusion
created by the dizzying world it bounded ... ('The Aleph', 13)

We must have a place to start, to begin reading the context. However
much the formative space of Los Angeles may be global (or perhaps
Mandelbrotian, constructed in zig-zagging nests of fractals), it must be
reduced to a more familiar and localized geometry to be seen. Appropriately enough, just such a reductionist mapping has popularly
presented itself. It is defined by an embracing circle drawn sixty miles
(about a hundred kilometres) out from a central point located in the
downtown core of the City of Los Angeles. Whether the precise central
point is City Hall or perhaps one of the more recently erected corporate
towers, I do not know. But I prefer the monumental twenty-eight storey
City Hall, up to the 1920s the only erection in the entire region allowed
to surpass the allegedly earthquake-proofing 150-foot height limitation.
It is an impressive punctuation point, capped by an interpretation of the
Mausoleum of Halicarnassus, wrapped around a Byzantine rotunda, and
etched with this infatuating inscription: 'THE CITY CAME INTO
BEING TO PRESERVE LIFE, IT EXISTS FOR THE GOOD LIFE'.
Significantly, City Hall sits at the corner of Temple and Spring Streets.

The Sixty-Mile Circle, so inscribed, covers the thinly sprawling 'built-
up' area of five counties, a population of more than 12 million individuals, at least 132 incorporated cities and, it is claimed, the greatest
concentration of technocratic expertise and militaristic imagination in
the USA. Its workers produce, when last estimated, a gross annual
output worth nearly $250 billion, more than the 800 million people of
India produce each year. This is certainly Greater Los Angeles, a dizzying world.

The determination of the Sixty-Mile Circle is the product of the
largest bank headquartered within its bounds, a bank potently named by
connecting together two definitive pillars of the circumscribed economy:
'security' and 'pacific'.[1] How ironic, indeed oxymoronic, is the combination of these two words, security and pacific. The first is redolent of the
lethal arsenal emanating from the Sixty-Mile Circle's technicians and
scientists, surely today the most powerful assemblage of weapon-making

---

1. At least eight editions of a pamphlet on 'The Sixty-Mile Circle' have been published
by the Economics Department of the Security Pacific National Bank, the first broadsheet
version appearing nearly twenty years ago. The 1981 edition aimed to celebrate the Los
Angeles Bicentennial. The edition I refer to, for 1984, advertises Security Pacific's support
of the Olympic Games.

expertise ever grounded into one place. In contrast, the second signals peacefulness, tranquillity, moderation, amity, concord. Holocaust attached to halcyon, another of the many simultaneous contraries, interposed opposites, which epitomize Los Angeles and help to explain why conventional categorical logic can never hope to capture its historical and geographic signification. One must return again and again to these simultaneous contraries to depict Los Angeles.

*Circumspection*

Securing the Pacific rim has been the manifest destiny of Los Angeles, a theme which defines its sprawling urbanization perhaps more than any other analytical construct. Efforts to secure the Pacific signpost the history of Los Angeles from its smoky inception as El Pueblo de Nuestra Señora la Reina de Los Angeles de Porciuncula in 1781, through its heated competition for commercial and financial hegemony with San Francisco, to the unfolding sequence of Pacific wars that has marked the past forty-five years of the American century. It is not always easy to see the imprint of this imperial history on the material landscape, but an imaginative cruise directly above the contemporary circumference of the Sixty-Mile Circle can be unusually revealing. Figure 11 will help to find the way.

The Circle cuts the south coast at the border between Orange and San Diego Counties, near one of the key checkpoints regularly set up to intercept the northward flow of undocumented migrants, and not far from the San Clemente 'White House' of Richard Nixon and the fitful SONGS of the San Onofre Nuclear Generating Station. The first rampart to watch, however, is Camp Pendleton Marine Corps Base, the largest military base in California in terms of personnel, the freed spouses of whom have helped to build a growing high-technology complex in northern San Diego County. After cruising over the moors of Camp Pendleton, the Cleveland National Forest, and the vital Colorado River Aqueduct draining in from the east[2] we can land directly in Rampart #2, March Air Force Base, adjacent to the city of Riverside. The insides of March are a ready outpost for the roaming Strategic Air Command.

Another quick hop over Sunnymead, the Box Spring Mountains, and Redlands takes us to Rampart #3, Norton Air Force Base, next to the city of San Bernardino and just south of the almost empty San Manuel Indian Reservation. The guide books tell us that the primary

---

2. The imperial history of the watering of Los Angeles is a key part of the growth of Southern California, but it cannot be treated here.

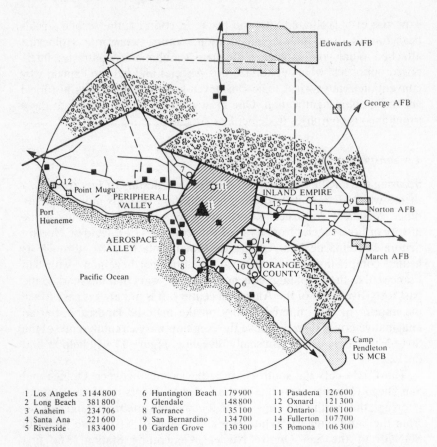

| | | | | | | | | |
|---|---|---|---|---|---|---|---|---|
| 1 | Los Angeles | 3144800 | 6 | Huntington Beach | 179900 | 11 | Pasadena | 126600 |
| 2 | Long Beach | 381800 | 7 | Glendale | 148800 | 12 | Oxnard | 121300 |
| 3 | Anaheim | 234706 | 8 | Torrance | 135100 | 13 | Ontario | 108100 |
| 4 | Santa Ana | 221600 | 9 | San Bernardino | 134700 | 14 | Fullerton | 107700 |
| 5 | Riverside | 183400 | 10 | Garden Grove | 130300 | 15 | Pomona | 106300 |

*Figure 11*   A view of the outer spaces of Los Angeles. The urban core is outlined in the shape of a pentagon, with the Central City denoted by the black triangle. The major military bases on the perimeter of the Sixty-Mile Circle are identified and the black squares are the sites of the largest defence contractors in the region. Also shown are county boundaries, the freeway system outside the central pentagon, and the location of all cities with more than 100,000 inhabitants (small open circles)

mission of Norton is military airlifts, just in case. To move on we must rise still higher to pass over the ski-sloped peaks of the San Bernardino Mountains and National Forest, through Cajon Pass and passing the old Santa Fe Trail, into the picturesque Mojave Desert. Near Victorville is Rampart #4, George Air Force Base, specializing in air defence and interception. Almost the same distance away – our stops seem remarkably evenly spaced thus far – takes us by dry Mirage Lake to the giant

Edwards Air Force Base, Rampart #5, site of NASA and USAF research and development activities and a primary landing field for unexploded Space Shuttles. Stretching off to the south is an important aerospace corridor through Lancaster, to Palmdale Airport and Air Force Plant 42, which serves Edwards's key historical function as testing ground for advanced fighters and bombers. People who live around here call it Canyon Country and many want it broken off from the County of Los Angeles down below.

The next leg is longer and more serene: over the Antelope Valley and the Los Angeles Aqueduct (tapping the Los Angeles-owned segments of the life-giving but rapidly dying Owens River Valley two hundred miles further away); across Interstate 5 (the main freeway corridor to the north), a long stretch of Los Padres National Forest and the Wild Condor Refuge,[3] to the idyll-ized town of Ojai (site for the filming of 'Lost Horizon'), and then to the Pacific again at the Mission of San Buenaventura, in Ventura County. A few miles away (the Sixty-Mile Circle actually cut right through the others) is Rampart #6, a complex consisting of a now inactive Air Force Base at Oxnard, the Naval Construction Battalion Center of Port Hueneme, and, far above all, the longsighted Naval Air Missile Center at Point Mugu. If we wished, we could complete the full circle of coincidence over the Pacific, picking up almost directly below us the US Naval Facilities on San Nicolas and San Clemente Islands. These islands rarely appear on maps of Los Angeles and they remain invisible on ours.

It is startling how much of the circumference is owned and preserved by the Federal Government in one way or another. Premeditation may be impossible to ascribe, but post-meditation on the circumscriptive federal presence is certainly in order.

*Enclosures*

What in the world lies behind this Herculean wall? What appears to need such formidable protection? In essence, we return to the same question with which we began: What is this place? There is, of course, that far-reaching Dream Machine and its launching pads, transmitting visual images and evocative sounds of that 'good life' announced on the facade of City Hall. But the 'entertainment' industry is itself a facade and, significant though it may be, there is much more being screened behind it, much more that has developed within the Sixty-Mile Circle that demands to be protected and preserved.

---

3. The last remaining condors were recently removed to zoos after lead-poisoning threatened their extinction in the 'wild'.

If there has emerged a compelling focus to the recent academic literature on Los Angeles, it is the discovery of extraordinary industrial production, a eureka so contrary to popular perceptions of Los Angeles that its explorers are often compelled to exaggerate to keep their lines of vision sufficiently open and clear against external obfuscations. Yet it is no exaggeration to claim that the Sixty-Mile Circle contains the premier industrial growth pole of the twentieth century, at least within the advanced capitalist countries. Oil, orange groves, films and flying set the scene at the beginning of the century and tend to remain fixed in many contemporary images of industrious, but not industrial, Los Angeles. Since 1930, however, Los Angeles has probably led all other major metropolitan areas in the USA, decade by decade, in the accumulation of new manufacturing employment.

For many, industrial Los Angeles nevertheless remains a contradiction in terms. When a colleague at UCLA (University of California, Los Angeles) began his explorations of the industrial geography of Los Angeles, his appeal to a prominent national scientific funding agency brought back a confidential referee's report (an economist, it appeared) proclaiming the absurdity of studying such a fanciful subject, something akin to examining wheat farming in Long Island. Fortunately sounder minds prevailed and the research progressed in exemplary fashion. Further evidence of the apparent invisibility of industrial production in Los Angeles came at about the same time from *Forbes* magazine, that self-proclaimed sourcebook for knowing capitalists (who should know better). In 1984, *Forbes* published a map identifying the major centres of high technology development in the USA. Cartographic attention was properly drawn to the Silicon Valley and the Route 128 axis around Boston, but all of Southern California was left conspicuously blank! Apparently invisible, hidden from view, was not only one of the historical source-regions for advanced technology in aerospace and electronics, but also what may well be the largest concentration of high technology industry and employment in the country (if not the world), the foremost silicon landscape, a region that has added over the past fifteen years a high-technology employment pool roughly equivalent to that of the whole image-fixing Silicon Valley of Santa Clara County to the north.

Still partially hidden behind this revelation are the primary generative agencies and the intricate processes producing this pre-eminent production complex. One key link, however, is abundantly clear. In the past half century, no other area has been so pumped with federal money as Los Angeles, via the Department of Defense to be sure, but also through numerous federal programmes subsidizing suburban consumption (suburbsidizing?) and the development of housing, transportation and water delivery systems. From the last Great Depression to the present, Los Angeles has been the prototypical Keynesian state-city, a federalized metro-sea of state-rescued capitalism enjoying its place in the sunbelt, demonstrating decade by decade its redoubtable ability to go first and multiply the public seed money invested in its promising economic landscape.[4] No wonder it remains so protected. In it are embedded many of the crown jewels of advanced industrial capitalism.

If anything, the federal flow is accelerating under the aegis of the military Keynesianism of the Reagan administration and the permanent arms economy of the Warfare State. At Hughes Aircraft Company in El Segundo, engineers have already used some of its $60 million in prime 'Star Wars' contracts to mock up a giant infra-red sensor so acute that it can pick up the warmth of a human body at a distance of a thousand miles in space, part of their experimentation with 'kinetic' weapons systems. Nearby, TRW Inc. ($84 million) and Rockwell International's Rocketdyne division ($32 million) competitively search for more powerful space lasers, capable, it seems, of incinerating whole cities if necessary, under such project code-names as Miracl, Alpha and Rachel. Research houses such as the Rand Corporation, just to the north in Santa Monica, jockey for more strategic positions, eager to claim part of what could potentially reach a total of $1.5 trillion if not stopped in time (Sanger, 1985). Today, it is not only the space of the Pacific that is being secured and watched over from inside the Sixty-Mile Circle.

*Outer spaces*

The effulgent Star Wars colony currently blooming around Los Angeles International Airport (LAX) is part of a much larger outer city which has taken shape along the Pacific slope of Los Angeles County. In the context of this landscape, through the story-line of the aerospace industry, can be read the explosive history and geography of the National Security State and what Mike Davis (1984) has called the 'Californianization of Late-Imperial America'.

If there is a single birthplace for this Californianization, it can be found at old Douglas Field in Santa Monica, today close by an important transit-point for President Reagan's frequent West Coast trips. From this spot fifty years ago the first DC-3 took off to begin a career of military accomplishment in war after war after war. Spinning off in its tracks has been an intricate tracery of links, from defence- and space-

---

4. The federalization of the Sixty-Mile Circle still remains poorly studied, but in this process are the forceful clues necessary for understanding the uneven regional development of the entire United States, Sunbelt and Frostbelt included.

related expenditures on research and development and the associated formation of the aerospace industry upon the base of civilian aircraft manufacturing; to the piggy-backed instigation of computerized electronics and modern information-processing technology, meshing with an ancillary network of suppliers and demanders of goods and services that stretches out to virtually every sector of the contemporary economy and society.[5] Over half a million people now live in this 'Aerospace Alley', as it has come to be called. During working hours, perhaps 800,000 are present to sustain its global pre-eminence. Untold millions more lie within its extended orbit.

Attached around the axes of production are the representative locales of the industrialized outer city: the busy international airport; corridors filled with new office buildings, hotels, and global shopping malls; neatly packaged playgrounds and leisure villages; specialized and master-planned residential communities for the high technocracy; armed and guarded housing estates for top professionals and executives; residual communities of low-pay service workers living in overpriced homes; and the accessible enclaves and ghettoes which provide dependable flows of the cheapest labour power to the bottom bulge of the bimodal local labour market. The LAX – City compage reproduces the segmentation and segregation of the inner city based on race, class, and ethnicity, but manages to break it down still further to fragment residential communities according to specific occupational categories, household composition, and a broad range of individual attributes, affinities, desired lifestyles and moods.

This extraordinary differentiation, fragmentation, and social control over specialized pools of labour is expensive. Housing prices and rental costs in the outer city are easily among the highest in the country and the provision of appropriate housing increasingly absorbs the energy not only of the army of real estate agents but of local corporate and community planners as well, often at the expense of long-time residents fighting to maintain their foothold in 'preferred' locations. From the give and take of this competition have emerged peculiarly intensified urban landscapes. Along the shores of the South Bay, for example, part of

---

5. In 1965, it was estimated in a Bank of America study that nearly 43 per cent of total manufacturing employment in Los Angeles and Orange counties was linked to defence and space expenditures. Some percentages by sector included paperboard containers and boxes (12 per cent), fabricated rubber (36 per cent), computing machines (54 per cent), photographic equipment (69 per cent), screw-machines (70 per cent) and machine shop jobbing and repairs (78 per cent). By 1983, almost half the manufacturing jobs in Los Angeles County were related directly or indirectly to the aerospace industry and nearly half of these aerospace workers were employed on military projects (Scheer, 1983). There is no reason to believe these figures have changed significantly since 1983.

what Rayner Banham (1971) once called 'Surfurbia', there has developed the largest and most homogenous residential enclave of scientists and engineers in the world (see Figure 8). Coincidentally, this beach-head of the high technocracy is also one of the most formidable racial redoubts in the region. Although just a few miles away, across the fortifying boundary of the San Diego freeway, is the edge of the largest and most tightly segregated concentration of Blacks west of Chicago, the sun-belted beach communities stretching south from the airport have remained almost 100 per cent white (Mate, 1982). The worldview from this highly engineered environment is stereotypically caricatured in Figure 12.

The Sixty-Mile Circle is ringed with a series of these outer cities at varying stages of development, each a laboratory for exploring the contemporaneity of capitalist urbanization. At least two are combined in Orange County, seamlessly webbed together into the largest and probably fastest growing outer city complex in the country (world?). The key nucleus here is the industrial complex embedded in the land empire of the Irvine Company, which owns one sixth of the entire county. Arrayed around it is a remarkable accretion of masterplanned new towns which paradigmatically evince the global cultural aspirations of the outer city imposed atop local visions of the experimental community of tomorrow.

Illustratively, the new town of Mission Viejo (never mind the bilingual pun) is partially blocked out to recreate the places and people of Cervantes's Spain and other quixotic intimations of the Mediterranean. Simultaneously, its ordered environment specifically appeals to Olympian dreams. Stacked with the most modern facilities and trainers, Mission Viejo has attracted an élite of sport-minded parents and accommodating children. The prowess of determined local athletes was sufficient for Mission Viejo to have finished ahead of 133 of the 140 countries competing in the 1984 Olympic Games in the number of medals received. Advertised as 'The California Promise' by its developer, currently the Philip Morris Company, Mission Viejo coughs up enticing portions of the American Dream to the chosen few. As one compromising resident described it, 'You must be happy, you must be well rounded and you must have children who do a lot of things. If you don't jog or walk or bike, people wonder if you have diabetes or some other disabling disease' (Landsbaum and Evans, 1984).

The Orange County complex has also been the focus for detailed research into the high technology industrial agglomerations that have been recentralizing the urban fabric of the Los Angeles region and inducing the fluorescence of masterplanned new towns. This pioneering work has helped us see more clearly the transactional web of indus-

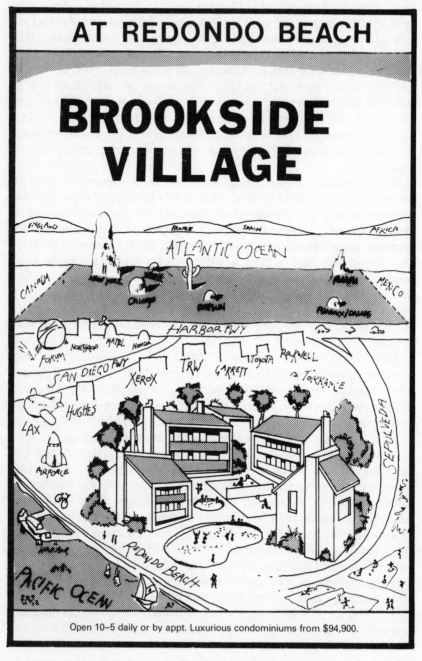

*Figure 12*   Worldview from Redondo Beach

trial linkages that draws out and geographically clusters specialized networks of firms, feeds off the flow of federal contracts, and spills over to precipitate a supportive local space economy (Scott, 1986). What has been provided is a revealing glimpse into the generative processes behind the urbanization of Orange County and, through this window, into the deeper historical interplay between industrialization and urbanization that has defined the development of the capitalist city wherever it is found.

There are other outer cities fringing the older pentagonal urban core outlined in Figure 11. One has taken shape in the Ventura Corridor through the west San Fernando Valley into Ventura County (now being called the 'Peripheral Valley', with its primary cores in 'Gallium Gulch' and the Chatsworth area).[6] Another is being promoted (although not yet in place) in the 'Inland Empire' stretching eastward from Pomona (General Dynamics is there) through Ontario (with Lockheed and a growing International Airport and Free Trade Zone) to the county seats of San Bernardino and Riverside, hard by their military ramparts. The inland empire, however, is still more of an anticipatory outer city, cruelly packed with new housing estates that automaniacally lure families ever further away from their places of work in Los Angeles and Orange counties, a truly transitory landscape.

Inland empirics aside, these new territorial complexes seem to be turning the industrial city inside-out, recentering the urban to transform the metropolitan periphery into the core region of advanced industrial production. Decentralization from the inner city has been taking place selectively for at least a century all over the world, but only recently has the peripheral condensation become sufficiently dense to challenge the older urban cores as centres of industrial production, employment nodality, and urbanism. This restructuring process is far from being completed but it is beginning to have some profound repercussions on the way we think about the city, on the words we use to describe urban forms and functions, and on the language of urban theory and analysis.

## Back to the Centre

I saw the teeming sea; I saw daybreak and nightfall; I saw the multitudes of America; I saw a silvery cobweb in the center of a black pyramid; I saw a

---

6. Gallium arsenide chips operate at higher frequencies and allegedly compute faster than silicon chips. Developed primarily for military use, they are expected by some to take an increasing share of the world semiconductor market in the future. 'Gallium Gulch' contains a cluster of recently formed companies experimenting with the new technology, all of which are headed by alumni of Rockwell International. (Goldstein, 1985)

splintered labyrinth ... I saw, close up, unending eyes watching themselves in me as in a mirror. ('The Aleph', 13)

To see more of Los Angeles, it is necessary to move away from the riveting periphery and return, literally and figuratively, to the centre of things to the still adhesive core of the urbanized landscape. In Los Angeles as in every city, the nodality of the centre defines and gives substance to the specificity of the urban, its distinctive social and spatial meaning. Urbanization and the spatial divisions of labour associated with it revolve around a socially constructed pattern of nodality and the power of the occupied centres both to cluster and disperse, to centralize and decentralize, to structure spatially all that is social and socially produced. Nodality situates and contextualizes urban society by giving material form to essential social relations. Only with a persistent centrality can there be outer cities and peripheral urbanization. Otherwise, there is no urban at all.

It is easy to overlook the tendential processes of urban structuration that emanate from the centre, especially in the postmodern capitalist landscape. Indeed, in contemporary societies the authoritative and allocative power of the urban centre is purposefully obscured or, alternatively, detached from place, ripped out of context, and given the stretched-out appearance of democratic ubiquity. In addition, as we have seen, the historical development of urbanization over the past century has been marked by a selective dispersal and decentralization, emptying the centre of many of the activities and populations which once aggregated densely around it. For some, this has signalled a negation of nodality, a submergence of the power of central places, perhaps even a Derridean deconstruction of all differences between the 'central' and the 'marginal'.

Yet the centres hold. Even as some things fall apart, dissipate, new nodalities form and old ones are reinforced. The specifying centrifuge is always spinning but the centripetal force of nodality never disappears. And it is the persistent residual of political power which continues to precipitate, specify, and contextualize the urban, making it all stick together. The first cities appeared with the simultaneous concentration of commanding symbolic forms, civic centres designed to announce, ceremonialize, administer, acculturate, discipline, and control. In and around the institutionalized locale of the citadel adhered people and their node-ordered social relations, creating a civil society and an accordingly built environment which were urbanized and regionalized through the interplay between two interactive processes, surveillance and adherence, looking out from and in towards a centre through the panoptic eye of power. To be urbanized still means to adhere, to be made an adherent, a believer in a specified collective ideology rooted in extensions of *polis* (politics, policy, polity, police) and *civitas* (civil, civic, citizen, civilian, civilization). In contrast, the population beyond the reach of the urban is comprised of *idiotes*, from the Greek root *idios*, meaning 'one's own, a private person', unlearned in the ways of the *polis* (a root akin to the Latin *sui*, 'of its own kind', with *generis*, 'constituting a class alone'). Thus to speak of the 'idiocy' of rural life or the urbanity of its opposition is primarily a statement of relative political socialization and spatialization, of the degree of adherence/separation in the collective social order, a social order hingeing on urban nodality.

To maintain adhesiveness, the civic centre has always served as a key surveillant node of the state, supervising locales of production, consumption and exchange. It still continues to do so, even after centuries of urban recomposition and restructuring, after waves of reagglomerative industrialization. It is not production or consumption or exchange in themselves that specifies the urban, but rather their collective surveillance, supervision and anticipated control within the power-filled context of nodality. In Foucauldian terms, cities are the convergent sites of (social) space, knowledge, and power, the headquarters of societal modes of regulation (from *regula* and *regere*, to rule; the root of our keyword: region).

This does not mean that a mechanical determinism is assigned to nodality in the specification of the urban. Adherence is a sticky notion and is not automatically enacted by location in an urbanized landscape; nor is it always awarely expressed in practical consciousness. Surveillance too is problematic, for it can exist without being embracingly effective – and can be embracingly effective without appearing to exist! There is thus always room for resistance, rejection, and redirection in the nonetheless structured field of urban locales, creating an active politics of spatiality, struggles for place, space, and position within the regionalized and nodal urban landscape. As a result, adherence and surveillance are unevenly developed in their geographical manifestation, their regionalization, their reactive regionalisms. Simultaneously, this patterned differentiation, this immediate superstructure of the urban spatial division of labour, becomes a critical arena in which the human geography of the city is shaped, in which spatialization takes place. It maps out an urban cartography of power and political praxis that is often hidden in idiographic (from *idios* again) histories and geographies.

## Signifying downtown

The downtown core of the City of Los Angeles, which the signs call 'Central City' is the agglomerative and symbolic nucleus of the Sixty-

Mile Circle, certainly the oldest but also the newest major node in the region. Given what is contained within the Circle, the physical size and appearance of downtown Los Angeles seem almost modest, even today after a period of enormous expansion. As usual, however, appearances can be deceptive.

Perhaps more than ever before, downtown serves in ways no other place can as a strategic vantage point, an urban panopticon counterposed to the encirclement of watchful military ramparts and defensive outer cities. Like the central well in Bentham's eminently utilitarian design for a circular prison, the original panopticon, downtown can be seen (when visibility permits) by each separate individual, from each territorial cell, within its orbit. Only from the advantageous outlook of the centre, however, can the surveillant eye see everyone collectively, disembedded but interconnected. Not surprisingly, from its origin, the central city has been an aggregation of overseers, a primary locale for social control, political administration, cultural codification, ideological surveillance, and the incumbent regionalization of its adherent hinterland.

Looking down and out from City Hall, the site is especially impressive to the observer. Immediately below and around is the largest concentration of government offices and bureaucracy in the country outside the federal capital district. To the east, over a pedestrian skyway, are City Hall East and City Hall South, relatively new civic additions enclosing a shopping mall, some murals, a children's museum, and the Triforium, a splashy sixty-foot fountain of water, light, and music entertaining the lunchtime masses. Just beyond is the imposing police administration building, Parker Center, hallowing the name of a former police chief of note. Looking further, outside the central well of downtown but within its eastern salient, one can see an area which houses 25 per cent of California's prison population, at least 12,000 inmates held in four jails designed to hold half that number. Included within this carceral wedge are the largest women's prison in the country (Sybil Brand) and the seventh largest men's prison (Men's Central). More enclosures are being insistently planned by the state to meet the rising demand.

On the south, along First Street, are the State Department of Transportation (CALTRANS) with its electronic wall maps monitoring the arterial freeways of the region, the California State Office Building, and the headquarters of the fourth estate, the monumental Times-Mirror building complex, which many have claimed houses the unofficial governing power of Los Angeles, the source of many stories that mirror the times and spaces of the city. Near the spatial sanctum of the *Los Angeles Times* is also St Vibiana's Cathedral, mother church to one of the largest Catholic archdioceses in the world (nearly four million strong) and controller of another estate of significant proportions. The Pope slept here, across the street from Skid Row missions temporarily closed so that he could not see all his adherents.

Looking westward now, toward the Pacific and the smog-hued sunsets which brilliantly paint the nightfalls of Los Angeles, is first the Criminal Courts Building, then the Hall of Records and Law Library, and next the huge Los Angeles County Courthouse and Hall of Administration, major seats of power for what is by far the country's largest county in total population (now over eight million). Standing across Grand Avenue is the most prominent cultural centre of Los Angeles, described by Unique Media Incorporated in their pictorial booster maps of downtown as 'the cultural crown of Southern California, reigning over orchestral music, vocal performance, opera, theatre and dance'. They add that the Music Center 'tops Bunker Hill like a contemporary Acropolis, one which has dominated civil cultural life since it was inaugurated in 1964[7]. Just beyond this cultural crown is the Department of Water and Power (surrounded by usually waterless fountains) and a multi-level extravaganza of freeway interchanges connecting with every corner of the Sixty-Mile Circle, a peak point of accessibility within the regional transportation network. On its edge, one of Japan's greatest architects has designed a Gateway Building to punctuate the teeming sea.

Along the northern flank is the Hall of Justice, the US Federal Courthouse, and the Federal Building, completing the ring of local, city, state and federal government authority which comprises the potent civic centre. Sitting more tranquilly just beyond, cut off by a swathe of freeway, is the preserved remains of the old civic centre, now part of El Pueblo de Los Angeles State Historical Park, additional testimony to the lasting power of the central place. Since the origins of Los Angeles the sites described have served as the political citadel, designed with other citadels to command, protect, socialize and dominate the surrounding urban population.

There is still another segment of the citadel – panopticon which cannot be overlooked. Its form and function may be more specific to the contemporary capitalist city but its mercantile roots entwine historically with the citadels of all urbanized societies. Today, it has become the acknowledged symbol of the urbanity of Los Angeles, the visual evidence of the successful 'search for a city' by the surrounding sea of

---

7. Colourful pictorial maps, so convenient for the exaggerated representation of presences and absences, seem to be multiplying at an unusually rapid pace all over Los Angeles, quietly erasing the unsightly, distorting spatial relations for effect and calling attention to the fantastic and the most merchandisable.

suburbs. This skylined sight contains the bunched castles and cathedrals of corporate power, the gleaming new 'central business district' of the 'central city', pinned next to its ageing predecessor just to the east. Here too the LA-leph's unending eyes are kept open and reflective, reaching out to and mirroring global spheres of influence, localizing the world that is within its reach.

Nearly all the landmarks of the new LA CBD have been built over the past fifteen years and flashily signify the consolidation of Los Angeles as a world city. Now more than half the major properties are in part or wholly foreign owned, although much of this landed presence is shielded from view. The most visible wardens are the banks which light up their logos atop the highest towers: Security Pacific (there again), First Interstate, Bank of America (co-owner of the sleek-black Arco Towers before their recent purchase by the Japanese), Crocker, Union, Wells Fargo, Citicorp (billing itself as 'the newest city in town'). Reading the skyline one sees the usual corporate panorama: large insurance companies (Manulife, Transamerica, Prudential), IBM and major oil companies, the real estate giant Coldwell Banker, the new offices of the Pacific Stock Exchange, all serving as attachment points for silvery webs of financial and commercial transactions extending practically everwhere on earth.

The two poles of the citadel, political and economic, connect physically through the condominium towers of renewed Bunker Hill but 'interface' less overtly in the planning apparatus of the local state. Contrary to popular opinion, Los Angeles is a tightly planned and plotted urban environment, especially with regard to the social and spatial divisions of labour necessary to sustain its pre-eminent industrialization and consumerism. Planning choreographs Los Angeles through the fungible movements of the zoning game and the flexible staging of supportive community participation (when there are communities to be found), a dance filled with honourable intent, dedicated expertise, and selective beneficence. It has excelled, however, as an ambivalent but nonetheless enriching pipeline and place-maker to the domestic and foreign developers of Los Angeles, using its influential reach to prepare the groundwork and facilitate the selling of specialized locations and populations to suit the needs of the most powerful organizers of the urban space-economy.[8]

---

8. Investigative reports of the political corruptibility of the planning process surface repeatedly in Los Angeles, with relatively little effect, for what is exposed is characteristically accepted as normal (if not normative) by the prominent practitioners. Two particularly thorough analyses appeared in 1985: Tony Castro, 'L.A. Inc', a three-part series in the *Herald Examiner* ('How politics built downtown', 10 March; 'Exercising political clout atop Bunker Hill', 11 March; and 'Critics claim CRA bulldozes over wishes of poor and

Although conspiracy and corruption can be easily found, the planned and packaged selling of Los Angeles usually follows a more mundane rhythm played to the legitimizing beat of dull and thumping market forces. In the created spaces which surround the twin citadels of Los Angeles, the beat has drummed with a particularly insistent and mesmerizing effect. Through a historic act of preservation and renewal, there now exists around downtown a deceptively harmonized showcase of ethni-cities and specialized economic enclaves which play key roles, albeit somewhat noisily at times, in the contemporary redevelopment and internationalization of Los Angeles. Primarily responsible for this packaged and planned production of the inner city is the Community Redevelopment Agency, probably the leading public entrepreneur of the Sixty-Mile Circle.[9]

There is a dazzling array of sites in this compartmentalized corona of the inner city: the Vietnamese shops and Hong Kong housing of a redeveloping Chinatown; the Big Tokyo financed modernization of old Little Tokyo's still resisting remains; the induced pseudo-SoHo of artists' lofts and galleries hovering near the exhibitions of the 'Temporary Contemporary' art warehouse; the protected remains of El Pueblo along Calmexified Olvera Street and in the renewed Old Plaza; the strangely anachronistic wholesale markets for produce and flowers and jewellery growing bigger while other downtowns displace their equivalents; the foetid sweatshops and bustling merchandise marts of the booming garment district; the Latino retail festival along pedestrian-packed Broadway (another preserved zone and inch-for-inch probably the most profitable shopping street in the region); the capital site of urban homelessness in the CRA-gilded skid row district; the enormous muralled *barrio* stretching eastward to the still unincorporated East Los Angeles; the de-industrializing and virtually resident-less wholesaling City of Vernon to the south filled with chickens and pigs awaiting their slaughter; the Central American and Mexican communities of Pico – Union and Alvarado abutting the high-rises on the west; the obtrusive oil wells and aggressive graffiti in the backyards of predominantly immigrant Temple–Beaudry progressively being eaten away by the spread of Central City West (now being called 'The Left Bank' of downtown); the

powerless', 12 March); and Ron Curran and Lewis MacAdams, 'The selling of L.A. County', *L.A. Weekly*, 22–28 November. Nothing comparable appeared in the *Los Angeles Times*.

9. The CRA is a California state-legislated agency directly responsible to the Los Angeles City Council. It functions publicly in downtown Los Angeles in facilitative ways which resemble the masterplanning operations of the Irvine Company in its private domains of Orange County.

intentionally yuppifying South Park redevelopment zone hard by the slightly seedy Convention Center; the revenue-milked towers and fortresses of Bunker Hill; the resplendently gentrified pocket of 'Victorian' homes in old Angelino Heights overlooking the citadel; the massive new Koreatown pushing out west and south against the edge of Black Los Angeles; the Filipino pockets to the north-west still un-coalesced into a 'town' of their own; and so much more: a constellation of Foucauldian heterotopias 'capable of juxtaposing in a single real place several spaces, several sites that are in themselves incompatible' but 'function in relation to all the space that remains'.

What stands out from a hard look at the inner city seems almost like an obverse (and perverse) reflection of the outer city, an agglomerative complex of dilapidated and overcrowded housing, low technology work-shops, relics and residuals of an older urbanization, a sprinkling of niches for recentred professionals and supervisors, and, above all, the largest concentration of cheap, culturally-splintered/occupationally-manipulable Third World immigrant labour to be found so tangibly available in any First World urban region. Here in this colonial corona is another of the crown jewels of Los Angeles, carefully watched over, artfully maintained and reproduced to service the continued development of the manufactured region.

The extent and persistence of agglomerated power and ever-watchful eyes in downtown Los Angeles cannot be ignored by either captive participants or outside observers. The industrialization of the urban peri-phery may be turning the space economy of the region inside-out, but the old centre is more than holding its own as the pre-eminent political and economic citadel. Peripheral visions are thus not enough when look-ing at Los Angeles. To conclude this spiralling tour around the power-filled central city, it may be useful to turn back to Giddens's observations on the structured and structuring landscapes of modern capitalism.

The distinctive structural principle of the class societies of modern capitalism is to be found in the disembedding, yet interconnecting, of state and economic institutions. The tremendous economic power generated by the harnessing of allocative resources to a generic tendency towards technical improvement is matched by an enormous expansion in the administrative 'reach' of the state. Surveillance – the coding of information relevant to the administration of subject populations, plus the direct supervision by officials and administrators of all sorts – becomes a key mechanism furthering a breaking away of system from social integration. Traditional practices are dispersed (without, of course, disappearing altogether) under the impact of the penetration of day-to-day life by codified administrative procedures. The locales which provide the settings for interaction in situations of co-presence [the basis for social integration] undergo a major set of transmutations. The old city-countryside relation is replaced by a sprawling expansion of a manufactured or 'created er.vironment'. (1984, 183–84)

Here we have another definition of spatial planning, another indication of the instrumentality of space and power, another example of spatialization.

*Lateral extensions*

Radiating from the specifying nodality of the central city are the hypo-thesized pathways of traditional urban theory, the transects of eagerly anticipated symmetries and salience which have absorbed so much of the attention of older generations of urban theoreticians and empiricists. Formal models of urban morphology have conventionally begun with the assumption of a structuring central place organizing an adherent landscape into discoverable patterns of hinterland development and regionalization. The deeper sources of this structuring process are usually glossed over and its problematic historical geography is almost universally simplified, but the resultant surfaces of social geometry continue to be visible as geographical expressions of the crude order-liness induced by the effects of nodality. They too are part of the spatial-ization of social life, the extended specificity of the urban.

The most primitive urban geometry arises from the radial attenuation of land use 'intensity' around the centre to an outer edge, a reflection of the Thunian landscape that has become codified most figuratively in the irrepressible Two-Parameter Negative Exponential Population Density Gradient. The TPNEPDG, in part because of its nearly universal and monotonous exemplification, has obsessed urban theorists with its projectable objectivity and apparent explanatory powers. From the Urban Ecologists of the old Chicago School to the New Urban Econo-mists, and including all those who are convinced that geographical analysis naturally begins with the primal explanation of variegated popu-lation densities (the most bourgeois of analytical assumptions Marx claimed), the TPNEPDG has been the lodestar for a monocentric understanding of urbanism. And within its own limited bands of confi-dence, it works efficiently.

Population densities do mound up around the centres of cities, even in the polycentric archipelago of Los Angeles (where there may be several dozen such mounds, although the most pronounced still falls off from the central city). There is also an accompanying concentric resi-dential rhythm associated with the family life cycle and the relative premiums placed on access to the dense peaks versus the availability of living space in the sparseness of the valleys (at least for those who can afford such freedoms of choice). Land values (when they can be accurately calculated) and some job densities also tend to follow in diminishing peaks outwards from the centre, bringing back to mind those tented webs of the urban geography textbooks.

Adding direction to the decadence of distance reduces the Euclidian elegance of concentric gradations, and many of the most mathematical of urban geometricians have accordingly refused to follow this slightly unsettling path. But direction does indeed induce another fit by pointing out the emanation of fortuitous wedges or sectors starting from the centre. The sectoral wedges of Los Angeles are especially pronounced once you leave the inner circle around downtown.

The Wilshire Corridor, for example, extends the citadels of the central city almost twenty miles westwards to the Pacific, picking up several other prominent but smaller downtowns en route (the Miracle Mile that initiated this extension, Beverly Hills, Century City, Westwood, Brentwood, Santa Monica). Watching above it is an even lengthier wedge of the wealthiest residences, running with almost staggering homogeneities to the Pacific Palisades and the privatized beaches of Malibu, sprinkled with announcements of armed responsiveness and signs which say that 'trespassers will be shot'. Here are the hearths of the most vocal home-owners movements, arms raised to slow growth and preserve their putative neighbourhoods in the face of the encroaching, view-blocking, street-clogging, and *declassé* downtowns.

As if in counterbalance, on the other side of the tracks east of downtown is the salient containing the largest Latino *barrio* in Anglo-America, where many of those who might be shot are carefully barricaded in poverty. And there is at least one more prominent wedge, stretching southward from downtown to the twin ports of Los Angeles–Long Beach, still reputed to be one of the largest consistently industrial urban sectors in the world. This is the primary axis of Ruhral Los Angeles.

A third ecological order perturbs the geometrical neatness still further, punching holes into the monocentric gradients and wedges as a result of the territorial segregation of races and ethnicities. Segregation is so noisy that it overloads the conventional statistical methods of urban factorial ecology with scores of tiny but 'significant' eco-components. In Los Angeles, arguably the most segregated city in the country, these components are so numerous that they operate statistically to obscure the spatiality of social class relations deeply embedded in the zones and wedges of the urban landscape, as if they needed to be obscured any further.[10]

---

10. Compacted Black Los Angeles has been particularly perplexing. For many statistical years, it contained both some of the lowest and the highest median family income

These broad social geometries provide an attractive model of the urban geography of Los Angeles, but like most of the inherited overviews of formal urban theory they are seriously diverting and illusory. They mislead not because there is disagreement over their degree of fit – such regular empiricist arguments merely induce a temporary insensibility by forcing debate on to the usually sterile grounds of technical discourse. Instead, they deceive by involuting explanation, by the legerdemain of making the nodality of the urban explain itself through its mere existence, one outcome explaining another. Geographical covariance in the form of empirico-statistical regularity is elevated to causation and frozen into place without a history – and without a human geography which recognizes that the organization of space is a social product filled with politics and ideology, contradiction and struggle, comparable to the making of history. Empirical regularities are there to be found in the surface geometry of any city, including Los Angeles, but they are not explained in the discovery, as is so often assumed. Different routes and different roots must be explored to achieve a practical understanding and critical reading of urban landscapes. The illusions of empirical opaqueness must be shattered, along with the other disciplining effects of Modern Geography.

*Deconstruction*

Back in the centre, shining from its circular turrets of bronzed glass, stands the Bonaventure Hotel, an amazingly storeyed architectural symbol of the splintered labyrinth that stretches sixty miles around it.[11] Like many other Portman-teaus which dot the eyes of urban citadels in New York and San Francisco, Atlanta and Detroit, the Bonaventure has become a concentrated representation of the restructured spatiality of the late capitalist city: fragmented and fragmenting, homogeneous and homogenizing, divertingly packaged yet curiously incomprehensible, seemingly open in presenting itself to view but constantly pressing to enclose, to compartmentalize, to circumscribe, to incarcerate. Everything imaginable appears to be available in this micro-urb but real places are difficult to find, its spaces confuse an effective cognitive mapping, its

census tracts in the county. And before 1970, the highest density of Black residents in any census tract was found not in the demarcable ghetto but in south Santa Monica. Blacks were among the original founders of Los Angeles in 1781, but there are virtually no major studies of the historical geography of Black Los Angeles currently available.

11. The Westin Bonaventure, financed and owned by the Japanese, figures prominently (if not with the correct spelling) in Jameson's perceptive analysis of postmodernism (1985). See also the rejoinder by Davis (1985) and an essay on postmodern planning by Michael Dear (1986). These writings form part of the first round of a continuing debate on postmodernism in Los Angeles.

pastiche of superficial reflections bewilder co-ordination and encourage submission instead. Entry by land is forbidding to those who carelessly walk but entrance is nevertheless encouraged at many different levels, from the truly pedestrian skyways above to the bunkered inlets below. Once inside, however, it becomes daunting to get out again without bureaucratic assistance. In so many ways, its architecture recapitulates and reflects the sprawling manufactured spaces of Los Angeles.

There has been no conspiracy of design behind the building of the Bonaventure or the socially constructed spatiality of the New World Cities. Both designs have been conjunctural, reflecting the specifications and exigencies of time and place, of period and region. The Bonaventure both simulates the restructured landscape of Los Angeles and is simultaneously simulated by it. From this interpretive interplay of micro- and macro-simulations there emerges an alternative way of looking critically at the human geography of contemporary Los Angeles, of seeing it as a mesocosm of postmodernity.

From the centre to the periphery, in both inner and outer cities, the Sixty-Mile Circle today encloses a shattered metro-sea of fragmented yet homogenized communities, cultures, and economies confusingly arranged into a contingently ordered spatial division of labour and power. As is true for so much of the patterning of twentieth-century urbanization, Los Angeles both sets the historical pace and most vividly epitomizes the extremes of contemporary expression. Municipal boundary making and territorial incorporation, to take one illustrative example, has produced the most extraordinary crazy quilt of opportunism to be found in any metropolitan area. Tiny enclaves of county land and whole cities such as Beverly Hills, West Hollywood, Culver City, and Santa Monica pock-mark the 'Westside' bulk of the incorporated City of Los Angeles, while thin slivers of city land reach out like tentacles to grab on to the key seaside outlets of the port at San Pedro and Los Angeles International Airport.[12] Nearly half the population of the city, however, lives in the quintessentially suburban San Fernando Valley, one and a half million people who statistically are counted as a part of the central city of the Los Angeles-Long Beach Standard Metropolitan

---

12. Another outlet reached near LAX is the Hyperion Sewage Treatment Plant, expectorating from the City of Los Angeles a volume of waste equivalent to the fifth or sixth largest river to reach the ocean in California; and creating an increasingly poisoned foodchain reaching back into the population of its drainage basin. Over the past several years, there have been claims that Santa Monica Bay may have the highest levels of toxic chemicals along the West coast. Signs were posted to warn of the hazards of locally caught fish (especially the so aptly named croaker) and doctors warned many of their patients not to swim off certain beaches. The fault-lines in the garbage-chains of the region may ultimately prove more threatening than those more well-known cracks in the earth's surface.

Statistical Area. Few other places make such a definitive mockery of the standard classifications of urban, suburban, and exurban.

Over 130 other municipalities and scores of county-administered areas adhere loosely around the irregular City of Los Angeles in a dazzling, sprawling patchwork mosaic. Some have names which are startlingly self-explanatory. Where else can there be a City of Industry and a City of Commerce, so flagrantly commemorating the fractions of capital which guaranteed their incorporation. In other places, names casually try to recapture a romanticized history (as in the many new communities called Rancho something-or-other) or to ensconce the memory of alternative geographies (as in Venice, Naples, Hawaiian Gardens, Ontario, Manhattan Beach, Westminster). In naming, as in so many other contemporary urban processes, time and space, the 'once' and the 'there', are being increasingly played with and packaged to serve the needs of the here and the now, making the lived experience of the urban increasingly vicarious, screened through *simulacra*, those exact copies for which the real originals have been lost.

A recent clipping from the *Los Angeles Times* (Herbert, 1985) tells of the 433 signs which bestow identity within the hyperspace of the City of Los Angeles, described as 'A City Divided and Proud of It'. Hollywood, Wilshire Boulevard's Miracle Mile, and the Central City were among the first to get these community signs as part of a 'city identification program' organized by the Transportation Department. One of the newest signs, for what was proclaimed 'the city's newest community', recognizes the formation of 'Harbor Gateway' in the thin eight-mile long blue-collar area threading south to the harbour, the old Shoestring Strip where many of the 32,000 residents often forgot their ties to the city. One of the founders of the programme pondered its development: 'At first, in the early 1960s, the Traffic Department took the position that all the communities were part of Los Angeles and we didn't want cities within cities ... but we finally gave in. Philosophically it made sense. Los Angeles is huge. The city had to recognize that there were communities that needed identification.... What we tried to avoid was putting up signs at every intersection that had stores.' Ultimately, the city signs are described as 'A Reflection of Pride in the Suburbs'. Where are we then in this nominal and noumenal fantasyland?

For at least fifty years, Los Angeles has been defying conventional categorical description of the urban, of what is city and what is suburb, of what can be identified as community or neighbourhood, of what co-presence means in the elastic urban context. It has in effect been deconstructing the urban into a confusing collage of signs which advertise what are often little more than imaginary communities and outlandish representations of urban locality. I do not mean to say that there are no genuine neighbourhoods to be found in Los Angeles. Indeed, finding them through car-voyages of exploration has become a popular local pastime, especially for those who have become so isolated from propinquitous community in the repetitive sprawl of truly ordinary-looking landscapes that make up most of the region. But again the urban experience becomes increasingly vicarious, adding more layers of opaqueness to *l'espace vécu.*

Underneath this semiotic blanket[13] there remains an economic order, an instrumental nodal structure, an essentially exploitative spatial division of labour, and this spatially organized urban system has for the past half century been more continuously productive than almost any other in the world. But it has also been increasingly obscured from view, imaginatively mystified in an environment more specialized in the production of encompassing mystifications than practically any other you can name. As has so often been the case in the United States, this conservative deconstruction is accompanied by a numbing depoliticization of fundamental class and gender relations and conflicts. When all that is seen is so fragmented and filled with whimsy and pastiche, the hard edges of the capitalist, racist and patriarchal landscape seem to disappear, melt into air.

With exquisite irony, contemporary Los Angeles has come to resemble more than ever before a gigantic agglomeration of theme parks, a lifespace comprised of Disneyworlds. It is a realm divided into showcases of global village cultures and mimetic American landscapes, all-embracing shopping malls and crafty Main Streets, corporation-sponsored magic kingdoms, high-technology-based experimental prototype communities of tomorrow, attractively packaged places for rest and recreation all cleverly hiding the buzzing workstations and labour processes which help to keep it together. Like the original 'Happiest Place on Earth', the enclosed spaces are subtly but tightly controlled by invisible overseers despite the open appearance of fantastic freedoms of choice. The experience of living here can be extremely diverting and exceptionally enjoyable, especially for those who can afford to remain inside long enough to establish their own modes of transit and places to rest. And, of course, the enterprise has been enormously profitable over the years. After all, it was built on what began as relatively cheap land, has been sustained by a constantly replenishing army of even cheaper imported labour, is filled with the most modern technological gadgetry, enjoys extraordinary levels of

---

13. The root of semiotic and semiology is the Greek *semeion,* which means sign, mark, spot or *point in space.* You arrange to meet someone at a *semeion,* a particular place. The significance of this connection between semiotics and spatiality is too often forgotten.

protection and surveillance, and runs under the smooth aggression of the most efficient management systems, almost always capable of delivering what is promised just in time to be useful.

## Afterwords

O God! I could be bounded in a nutshell, and count myself a King of infinite space ...

(*Hamlet*, II, 2; first prescript to *The Aleph*)

But they will teach us that Eternity is the Standing still of the Present Time, a *Nunc-stans* (as the Schools call it); which neither they, nor any else understand, no more than they would a *Hic-stans* for an infinite greatness of Place.

(*Leviathan*, IV, 46; second prescript to *The Aleph*)

I have been looking at Los Angeles from many different points of view and each way of seeing assists in sorting out the interjacent medley of the subject landscape. The perspectives explored are purposeful, eclectic, fragmentary, incomplete, and frequently contradictory, but so too is Los Angeles and, indeed, the experienced historical geography of every urban landscape. Totalizing visions, attractive though they may be, can never capture all the meanings and significations of the urban when the landscape is critically read and envisioned as a fulsome geographical text. There are too many *auteurs* to identify, the *literalité* (materiality?) of the manufactured environment is too multilayered to be allowed to speak for itself, and the countervailing metaphors and metonyms frequently clash, like discordant symbols drowning out the underlying themes. More seriously, we still know too little about the descriptive grammar and syntax of human geographies, the phonemes and epistemes of spatial interpretation. We are constrained by language much more than we know, as Borges so knowingly admits: what we can see in Los Angeles and in the spatiality of social life is stubbornly simultaneous, but what we write down is successive, because language is successive. The task of comprehensive, holistic regional description may therefore be impossible, as may be the construction of a compleat historico-geographical materialism.

There is hope nonetheless. The critical and theoretical interpretation of geographical landscapes has recently expanded into realms that functionally had been spatially illiterate for most of the twentieth century. New and avid readers abound as never before, many are directly attuned to the specificity of the urban, and several have significantly turned their eyes to Los Angeles. Moreover, many practised readers of surface geographies have begun to see through the alternatively myopic and hypermetropic distortions of past perspectives to bring new insight to spatial analysis and social theory. Here too Los Angeles has attracted observant readers after a history of neglect and misapprehension, for it insistently presents itself as one of the most informative palimpsests and paradigms of twentieth-century urban-industrial development and popular consciousness.

As I have seen and said in various ways, everything seems to come together in Los Angeles, the totalizing Aleph. Its representations of spatiality and historicity are archetypes of vividness, simultaneity, and interconnection. They beckon inquiry at once into their telling uniqueness and, at the same time, into their assertive but cautionary generalizability. Not all can be understood, appearances as well as essences persistently deceive, and what is real cannot always be captured even in extraordinary language. But this makes the challenge more compelling, especially if once in a while one has the opportunity to take it all apart and reconstruct the context. The reassertion of space in critical social theory – and in critical political praxis – will depend upon a continued deconstruction of a still occlusive historicism and many additional voyages of exploration into the heterotopias of contemporary postmodern geographies.

# Bibliography

Aglietta, M. (1979) *A Theory of Capitalist Regulation: The U.S. Experience*, London: New Left Books.

Amin, S. (1974) *Accumulation on a World Scale: A Critique of the Theory of Under-development*, New York: Monthly Review.

Amin, S. (1973) *Unequal Exchange*, New York: Random House.

Anderson, J. (1980) 'Towards a Materialist Conception of Geography', *Geoforum* 11, 171–78.

Anderson, P. (1983) *In the Tracks of Historical Materialism*, London: Verso.

Anderson, P. (1980) *Arguments Within English Marxism*, London: Verso.

Anderson, P. (1976) *Considerations on Western Marxism*, London: Verso.

Bachelard, G. (1969) *The Poetics of Space*, Boston: Beacon; trans. by M. Jolas of *La Poétique de l'éspace*, Paris: PUF, 1957.

Bagnasco, A. (1977) *Le Tre Italie*, Bologna: Il Mulino.

Banham, R. (1971) *Los Angeles: The Architecture of the Four Ecologies*, New York: Harper and Row.

Berger, J. (1984) *And our faces, my heart, brief as photos*, New York, Pantheon Books.

Berger, J. (1980) *About Looking*, London: Writers and Readers Publishing Cooperative; New York: Pantheon Books.

Berger, J. (1974) *The Look of Things*, New York: The Viking Press.

Berger, J. (1972) *Ways of Seeing*, London and Harmondsworth: British Broadcasting Corporation and Penguin Books.

Bergson, H. (1910) *Time and Free Will: An Essay on the Immediate Data of Consciousness*, trans. by F. Pogson, London: G. Allen; New York: Macmillan.

Berman, M. (1982) *All That Is Solid Melts Into Air: The Experience of Modernity*, New York: Simon and Schuster; also 1983, London, Verso.

Bernstein, R. (ed.) (1985) *Habermas and Modernity*, Oxford: Polity Press.

Bhaskar, R. (1986) *Scientific Realism and Human Emancipation*, London: Verso.

Bhaskar, R. (1979) *The Possibility of Naturalism*, Brighton: Harvester Press.

Bhaskar, R. (1975) *A Realist Theory of Science*, Leeds: Alma.

Blaut, J. (1975) 'Imperialism: The Marxist Theory and its Evolution', *Antipode* 7, 1–19.

Bluestone, B., and Harrison, B. (1982) *The Deindustrialization of America*, New York: Basic Books.

Borges, J.L. (1971) 'The Aleph', *The Aleph and Other Stories: 1933–1969*, New York: Bantam Books, 3–17.

Brenner, R. (1977) 'The Origins of Capitalist Development: a Critique of Neo-Smithian Marxism', *New Left Review* 104, 25–92.

Brookfield, H. (1975) *Interdependent Development*, London: Methuen.

*249*

Buber, M. (1957) 'Distance and Relation', *Psychiatry* 20, 97–104.
Burgel, Gallia, Burgel G., and Dezes, M. (1987) 'An Interview with Henri Lefebvre', translated by E. Kofman, *Environment and Planning D: Society and Space* 5, 27–38.
Carney, J., Hudson, R., and Lewis, J. (eds) (1980) *Regions in Crisis*, London: Croom Helm.
Castells, M. (1985) 'High Technology, Economic Restructuring, and the Urban-regional Process in the United States', in Castells (ed.) *High Technology, Space, and Society*, Beverly Hills: Sage, 11–40.
Castells, M. (1983) *The City and the Grass Roots*, Berkeley and Los Angeles: University of California Press.
Castells, M. (1977) *The Urban Question*, London: Edward Arnold; trans. of *La Question urbaine* (1972) Paris: Maspero.
Castells, M. (1976) 'The wild city', *Kapitalistate* 4, 2–30.
Castells, M., and Godard, F. (1974) *Monopolville: l'entreprise, l'état, l'urbain*, Paris: Mouton.
Cochrane, A. (1987) 'What a Difference the Place Makes: the New Structuralism of Locality', *Antipode* 19, 354–63.
Cohen, G.A. (1978) *Karl Marx's Theory of History: A Defense*, Princeton, Princeton University Press.
Cooke, P. (1987) 'Clinical Inference and Geographic Theory', *Antipode* 19, 69–78.
Cooke, P. (1983) *Theories of Planning and Spatial Development*, London: Hutchinson.
Davis, M. (1987) '*Chinatown*, Part Two? The Internationalization of Downtown Los Angeles', *New Left Review* 164, 65–86.
Davis, M. (1985) 'Urban Renaissance and the Spirit of Postmodernism', *New Left Review* 151, 106–13.
Davis, M. (1984) 'The Political Economy of Late Imperial America', *New Left Review* 143, 6–38.
Dear, M. (1986) 'Postmodernism and Planning', *Environment and Planning D: Society and Space* 4, 367–84.
Dear, M., and Scott, A. (eds) (1981) *Urbanization and Urban Planning in Capitalist Society*, London and New York: Methuen.
Duncan, J., and Ley, D. (1982) 'Structural Marxism and Human Geography: A Critical Assessment', *Annals, Association of American Geographers* 72, 30–58.
Eagleton, T. (1986) *Against the Grain: Essays 1975–1985*, London: Verso.
Edel, M. (1977) 'Rent Theory and Working Class Strategy: Marx, George and the Urban Crisis', *The Review of Radical Political Economics* 9.
Eliade, M. (1959) *Cosmos and History*, New York: Harper.
Eliot Hurst, M. (1980) 'Geography, Social Science and Society: Towards a De-definition', *Australian Geographical Studies* 18, 3–21.
Emmanuel, A. (1972) *Unequal Exchange: A Study of the Imperialism of Trade*, New York: Modern Reader.
Eyles, J. (1981) 'Why Geography Cannot be Marxist: Towards an Understanding of Lived Experience', *Environment and Planning A* 12, 1371–88.
Fell, J. (1979) *Heidegger and Sartre: An Essay on Being and Place*, New York: Columbia University Press.
Forbes, D., and Rimmer, P. (eds) (1984) *Uneven Development and the Geographical Transfer of Value*, Canberra: The Australian National University Research School of Pacific Studies, Human Geography Monograph 16.
Forbes, D., and Thrift, N. (1984) 'Determination and Abstraction in Theories of the Articulation of Modes of Production', in Forbes and Rimmer (eds) 111–34.
Foster, H. (1983) *The Anti-Aesthetic*, Port Townsend, WA: Bay Press; also published 1985 as *Postmodern Culture*, London: Pluto.
Foucault, M. (1986) 'Of Other Spaces', *Diacritics* 16, 22–27 (translated from the French by Jay Miskowiec).
Foucault, M. (1980) 'Questions on Geography', in C. Gordon (ed.), *Power/Knowledge: Selected Interviews and Other Writings 1972–1977*, 63–77.

Foucault, M. (1978) *The History of Sexuality Volume 1: An Introduction*, translated by R. Hurley, New York: Pantheon.

Foucault, M. (1977) 'The Eye of Power', preface (*'L'œil de pouvoir'*) to J. Bentham, *Le Panoptique*, Paris: Belfond: reprinted in Gordon (ed.) *Power/Knowledge*.

Foucault, M. (1961) *Histoire de la folie a l'âge classique*, Paris: Gallimard, translated by R. Howard as *Madness and Civilization*, London: Tavistock.

Friedmann, J. (1972) 'A General Theory of Polarized Development', in N. Hansen (ed.) *Growth Centers in Regional Economic Development*, 82–107.

Friedmann, J., and Weaver, C. (1979) *Territory and Function*, Berkeley and Los Angeles: University of California; and London: Edward Arnold.

Friedmann, J., and Wolff, G. (1982) 'World City Formation: an Agenda for Research and Action', *International Journal of Urban and Regional Research* 6, 309–44.

Gibson, K., Graham, J., Shakow, D., and Ross, R. (1984) 'A Theoretical Approach to Capital and Labour Restructuring', in P. O'Keefe (ed.), 39–64.

Gibson, K., and Horvath, R. (1983) 'Aspects of a Theory of Transition within the Capitalist Mode of Production', *Environment and Planning D: Society and Space* 1, 121–38.

Giddens, A. (1984) *The Constitution of Society: Outline of the Theory of Structuration*, Cambridge: Polity Press; and Berkeley and Los Angeles: University of California Press.

Giddens, A. (1981) *A Contemporary Critique of Historical Materialism*, London and Basingstoke: Macmillan; and Berkeley and Los Angeles: University of California Press.

Giddens, A. (1979) *Central Problems in Social Theory*, London and Basingstoke; Macmillan; and Berkeley and Los Angeles: University of California Press.

Giddens, A. (1976) *New Rules of Sociological Method*, New York: Basic Books.

Goldstein, A. (1985) 'Southland Firms Race for Lead in High-tech Material', *Los Angeles Times* 6 August, part IV, 2, 6.

Goodman, R. (1979) *The Last Entrepreneurs: America's Regional Wars for Jobs and Dollars*, New York: Simon and Shuster.

Gordon, C. (ed.) (1980) *Power/Knowledge: Selected Interviews and Other Writings 1972–1977*, New York: Pantheon.

Gordon, D. (1978) 'Capitalist Development and the History of American Cities', in W. Tabb and L. Sawers (eds) *Marxism and the Metropolis*, New York: Oxford University Press.

Gordon, D. (1977) 'Class Struggle and the Stages of American Urban Development', in D. Perry and A. Watkins (eds) *The Rise of the Sunbelt Cities*, Beverly Hills: Sage.

Gregory, D. (forthcoming) *The Geographical Imagination*, London: Hutchinson.

Gregory, D. (1984) 'Space, Time and Politics in Social Theory: an Interview with Anthony Giddens', *Environment and Planning D: Society and Space* 2, 123–32.

Gregory, D. (1981) 'Human Agency and Human Geography', *Transactions of the IBG* 6, 1–18.

Gregory, D. (1978) *Ideology, Science and Human Geography*, London: Hutchinson.

Gregory, D., and Urry, J. (eds) (1985) *Social Relations and Spatial Structures*, London: Macmillan; and New York: St Martin's.

Gregson, N. (1987) 'The CURS Initiative: Some Further Comments', *Antipode* 19, 364–70.

Gross, D. (1981–1982) 'Space, Time, and Modern Culture', *Telos* 50, 59–78.

Hadjimichalis, C. (1986) *Uneven Development and Regionalism: State, Territory and Class in Southern Europe*, London: Croom Helm.

Hadjimichalis, C. (1984) 'The Geographical Transfer of Value: Notes on the Spatiality of Capitalism', *Environment and Planning D: Society and Space* 2, 329–45.

Hadjimichalis, C. (1980) *The Geographical Transfer of Value* Ph.D. Dissertation, Graduate School of Architecture and Urban Planning, University of California, Los Angeles.

Harloe, M. (ed.) (1976) *Captive Cities*, New York: John Wiley and Sons.

Harre, R. (1970) *The Principles of Scientific Thinking*, London: Macmillan.

Harre, R. (1979) *Social Being*, Oxford: Basil Blackwell.

Harre, R., and Madden, E. (1975) *Causal Powers*, Oxford: Basil Blackwell.

Harrison, B. (1984) 'Regional Restructuring and "Good Business Climates": the Economic Transformation of New England since World War II', in L. Sawers and W. Tabb (eds) *Sunbelt/Snowbelt: Urban Development and Regional Restructuring*, New York: Oxford University Press.

Hartshorne, R. (1959) *Perspective on the Nature of Geography*, Chicago: Rand McNally.

Hartshorne, R. (1939) *The Nature of Geography: A Critical Survey of Current Thought in the Light of the Past*, Lancaster, PA: Association of American Geographers.

Harvey, D. (1987) 'Flexible Accumulation Through Urbanisation: Reflections on "Postmodernism", in the American city', *Antipode* 19, 260–86.

Harvey, D. (1985a) *The Urbanization of Capital*, Baltimore: Johns Hopkins University Press; and Oxford: Basil Blackwell.

Harvey, D. (1985b) *Consciousness and the Urban Experience*, Baltimore: Johns Hopkins University Press; and Oxford: Basil Blackwell.

Harvey, D. (1985c) 'The Geopolitics of Capitalism', in Gregory and Urry (eds) *Social Relations and Spatial Structures*, 126–63.

Harvey, D. (1984) 'On the History and Present Condition of Geography: an Historical Materialist Manifesto', *Professional Geographer* 36, 1–11.

Harvey, D. (1982) *The Limits to Capital*, Oxford: Basil Blackwell; and Chicago: University of Chicago Press.

Harvey, D. (1981) 'The Spatial Fix: Hegel, von Thunen and Marx', *Antipode* 13, 1–12.

● Harvey, D. (1978) 'The Urban Process under Capitalism', *International Journal of Urban and Regional Research* 2, 101–31.

Harvey, D. (1977) 'Labor, Capital and Class Struggle Around the Built Environment in Advanced Capitalist Societies', *Politics and Society* 6, 265–95.

Harvey, D. (1975) 'The Geography of Capitalist Accumulation: A Reconstruction of Marxian Theory', *Antipode* 7, 9–21.

Harvey, D. (1973) *Social Justice and the City*, Baltimore: Johns Hopkins University Press; and London: Edward Arnold.

Harvey, D. (1969) *Explanation in Geography*, New York: St Martin's; and London: Edward Arnold.

Harvey, D. et al. (1987) 'Reconsidering Social Theory: a Debate', *Environment and Planning D: Society and Space* 5, 367–434.

Heidegger, M. (1962) *Being and Time*, Oxford: Basil Blackwell.

Herbert, R. (1985) 'LA – A City Divided and Proud of It', *Los Angeles Times* 9 December, part I, 1. 3. 30.

Hirsch, A. (1981) *The French New Left: An Intellectual History from Sartre to Gorz*, Boston: Southend.

Hirschman, A. (1958) *The Strategy of Economic Development*, New Haven: Yale University Press.

Hobsbawm, E.J. (1987) *The Age of Empire 1875–1914*, New York: Pantheon.

Hobsbawm, E.J. (1975) *The Age of Capital 1848–1875*, New York: Charles Scribner's Sons.

Hobsbawm, E.J. (1962) *The Age of Revolution 1789–1848*, New York: New American Library.

Hughes, H.S. (1958) *Consciousness and Society: The Reconstruction of European Social Thought 1890–1930*, New York: Knopf.

Hutcheon, L. (1987) 'Beginning to Theorize Postmodernism', *Textual Practice* 1, 10–31.

Jackson, K. (1985) *The Crabgrass Frontier: The Suburbanization of America*, New York: Oxford University Press.

Jameson, F. (1984) 'Postmodernism, or the Cultural Logic of Late Capitalism', *New Left Review* 146, 53–92.

Jameson, F. (1984) *Sartre: The Origins of a Style*, New York: Columbia University Press.

Jay, M. (1984) *Marxism and Totality: The Adventures of a Concept from Lukacs to Habermas*, Berkeley and Los Angeles: University of California Press.

Johnston, R.J., Gregory, D., and Smith, D.M. (1986) *The Dictionary of Human Geography*, Oxford: Basil Blackwell.

Keat, R., and Urry, J. (1982) *Social Theory as Science*, London: Routledge and Kegan Paul.
Kelly, M. (1982) *Modern French Marxism*, Oxford: Basil Blackwell; and Baltimore: Johns Hopkins University Press.
Kern, S. (1983) *The Culture of Time and Space 1880–1918*, Cambridge: Harvard University Press.
Kidron, M. (1974) *Capitalism and Theory*, London: Pluto.
Kopp, A. (1971) *Town and Revolution*, Paris: Brazillar.
Lacoste, Y. (1976) *La Geographie, ça sert, d'abord, à faire la guerre*, Paris: Maspero.
Landsbaum, M., and Evans, H. (1984) 'Mission Viejo: Winning is the Only Game in Town', *Los Angeles Times* 22 August, part I, 1, 26, 27.
Lash, S., and Urry, J. (1987) *The End of Organized Capitalism*, Cambridge: Polity Press.
Lefebvre, H. (1980) *Une pensée devenue monde: faut-il abandonner Marx?*, Paris: Fayard.
Lefebvre, H. (1976–78) *De l'état*, 4 vols, Paris: Union Générale d'Éditions.
Lefebvre, H. (1976a) *The Survival of Capitalism*, London: Allison and Busby.
Lefebvre, H. (1976b) 'Reflections on the Politics of Space', translated by M. Enders, *Antipode* 8, 30–37.
Lefebvre, H. (1975) *Le Temps des méprises*, Paris: Stock.
Lefebvre, H. (1974) *La Production de l'espace*, Paris: Anthropos.
Lefebvre, H. (1973) *La Survie du capitalisme*, Paris: Anthropos.
Lefebvre, H. (1972) *La Pensée marxiste et la ville*, Paris: Casterman.
Lefebvre, H. (1971) *Au-delà du structuralisme*, Paris: Editions Anthropos.
Lefebvre, H. (1970a) *La Révolution urbaine*, Paris: Gallimard.
Lefebvre, H. (1970b) *Manifeste différentialiste*, Paris: Gallimard.
Lefebvre, H. (1968a) *La Vie quotidienne dans le monde moderne*, Paris: Gallimard.
Lefebvre, H. (1968b) *Le Droit à la ville*, Paris: Anthropos.
Lefebvre, H. (1961) *Fondements d'une sociologie de quotidienneté*, Paris: L'Arche.
Lefebvre, H. (1946b, reissued 1958) *Critique de la vie quotidienne*, Paris: L'Arche.
Lefebvre, H. (1946) *L'Existentialisme*, Paris: Éditions du Sagittaire.
Lefebvre, H., and Guterman, N. (1936) *La Conscience mystifiée*, Paris: Gallimard.
Lipietz, A. (1986) 'New Tendencies in the International Division of Labor: Regimes of Accumulation and Modes of Regulation', in Scott and Storper (eds), 16–40.
Lipietz, A. (1984a) 'Accumulation, crises et sorties de crise: quelques réflexions méthodologiques autour de la notion de "régulation"', Paris: CEDREMAP RP 8409.
Lipietz, A. (1984b) 'The Globalization of the General Crisis of Fordism', Paris: CEDREMAP RP 8413.
Lipietz, A. (1980) 'The Structuration of Space, the Problem of Land, and Spatial Policy', in Carney et al. (eds), 60–75.
Lipietz, A. (1977) *Le Capital et son espace*, Paris: Maspero.
Lowith, K. (1949) *Meaning in History*, Chicago: University of Chicago Press.
Lynch, K. (1960) *The Image of the City*, Cambridge, MA; and London, MIT Press.
Lyotard, J-F. (1986) *The Postmodern Condition: A Report on Knowledge*, Manchester: Manchester University Press.
Mackinder, H. (1919) *Democratic Ideals and Reality: A Study in the Politics of Reconstruction*, London: Constable.
Mackinder, H. (1904) 'The Geographical Pivot of History', *Geographical Journal* 23, 421–37.
Mandel, E. (1980) *Long Waves in Capitalist Development: The Marxist Interpretation*, Cambridge: Cambridge University Press.
Mandel, E. (1978) *The Second Slump*, London: Verso.
Mandel, E. (1976) 'Capitalism and Regional Disparities', *Southwest Economy and Society* 1, 41–47.
Mandel, E. (1975) *Late Capitalism*, London: Verso.
Mandel, E. (1968) *Marxist Economic Theory*, New York: Monthly Review.
Mandel, E. (1963) 'The Dialectic of Class and Region in Belgium', *New Left Review* 20, 5–31.

Mann, E. (1987) *Taking On General Motors: A Case Study of the UAW Campaign to Keep GM Van Nuys Open*, Los Angeles: UCLA Institute of Industrial Relations, Center for Labour Research and Education.

Mann, M. (1986) *The Sources of Social Power Volume 1: A History of Power from the Beginning to A.D. 1760*, Cambridge, Cambridge University Press.

Markusen, A. (1978) 'Regionalism and the Capitalist State: the Case of the United States', *Kapitalistate* 7, 39–62.

Martins, M. (1983) 'The Theory of Social Space in the Work of Henri Lefebvre', in Forrest, Henderson, and Williams (eds) *Urban Political Economy and Social Theory*, (Epping: Gower) 160–85.

Marx, K. (1973) *Grundrisse*, translated by M. Nicolaus, Harmondsworth: Penguin, in association with *New Left Review*.

Massey, D. (1984) *Spatial Divisions of Labour: Social Structures and the Geography of Production*, London and Basingstoke: Macmillan.

Massey, D. (1978) 'Regionalism: Some Current Issues', *Capital and Class* 6, 106–25.

Mate, K. (1982) '"For Whites Only": The South Bay Perfects Racism for the '80s', *LA Weekly* 6–12 August, 11ff.

May, J.A. (1970) *Kant's Concept of Geography and its Relation to Recent Geographical Thought*, Toronto: University of Toronto Press, Department of Geography Research Publications.

Miller, G. (1981) *Cities by Contract: The Politics of Municipal Incorporation*, Cambridge MA: MIT Press.

Mills, C.W. (1959) *The Sociological Imagination*, New York: Oxford University Press.

Molotch, H., and Logan, J. (1987) *Urban Fortunes: The Political Economy of Place*, Berkeley and Los Angeles: University of California Press.

Morales, R. (1986) 'The Los Angeles Automobile Industry in Historical Perspective', *Environment and Planning D: Society and Space* 4, 289–303.

Morales, R. (1983) 'Transitional Labor: Undocumented Workers in the Los Angeles Automobile Industry', *International Migration Review* 17, 570–96.

Morales, R., Azores, T., Purkey, R., and Ulgen, S. (1982) *The Use of Shift-Share Analysis in Studying the Los Angeles Economy 1962–1977*, Los Angeles, UCLA Graduate School of Architecture and Urban Planning Publications, Report 58.

Myrdal, G. (1957) *Rich Lands and Poor*, New York: Harper and Row.

*New German Critique* (1984), 'Modernity and Postmodernity, 33, 1–269.

Offe, C. (1985) *Disorganized Capitalism*, Cambridge MA: MIT Press.

O'Keefe, P. (ed.) (1984) *Regional Restructuring Under Advanced Capitalism*, Beckenham: Croom Helm.

Olsson, G. (1980) *Birds in Egg/Eggs in Bird*, New York: Methuen; and London: Pion.

Palloix, C. (1977) 'The Self Expansion of Capital on a World Scale', *The Review of Radical Political Economics* 9, 1–28.

Parson, D. (1985) *Urban Politics During the Cold War: Public Housing, Urban Renewal and Suburbanization in Los Angeles*, PhD Dissertation, Graduate School of Architecture and Urban Planning, University of California, Los Angeles.

Parson, D. (1982) 'The Development of Redevelopment: Public Housing and Urban Renewal in Los Angeles', *International Journal of Urban and Regional Research* 6, 393–413.

Peet, R. (1981) 'Spatial Dialectics and Marxist Geography', *Progress in Human Geography* 5, 105–10.

Peet, R. (ed.) (1977) *Radical Geography: Alternative Viewpoints on Contemporary Social Issues*, Chicago, Maaroufa; and London: Methuen.

Peet, R., and Thrift, N. (eds) (1988) *The New Models in Geography*, Hemel Hempstead: Allen and Unwin.

Perry, L., and Perry, R. (1963) *A History of the Los Angeles Labor Movement 1911–1941*, Berkeley and Los Angeles: University of California Press.

Pickles, J. (1985) *Phenomenology, Science and Geography*, Cambridge: Cambridge University Press.

Piore, M., and Sabel, C. (1984) *The Second Industrial Divide: Possibilities for Prosperity*, New York: Basic Books.

Plotkin, S. (1987) *Keep Out: The Struggle for Land Use Control*, Berkeley and Los Angeles: University of California Press.

Popper, K. (1957) *The Poverty of Historicism*, London and Boston: Routledge and Kegan Paul.

Poster, M. (1975) *Existential Marxism in Postwar France*, Princeton: Princeton University Press.

Poster, M. (1979) *Sartre's Marxism*, London: Pluto.

Poulantzas, N. (1978) *State, Power, Socialism*, London: Verso.

Pred, A. (1986) *Place, Practice and Structure: Social and Spatial Transformation in Southern Sweden 1750-1850*, Cambridge: Cambridge University Press.

Pred, A. (1984) 'Structuration, Biography Formation and Knowledge: Observations on Port Growth During the Late Mercantile Period', *Environment and Planning D: Society and Space* 2, 251-75.

Rabinow, P. (ed.) (1984) *The Foucault Reader*, New York: Pantheon Books.

Rabinow, P. (1984) 'Space, Knowledge, and Power', in P. Rabinow (ed.), *The Foucault Reader*, 239-56.

Rorty, R. (1980) *Philosophy and the Mirror of Nature*, Oxford: Basil Blackwell.

Roweis, S. (1975) 'Urban Planning in Early and Late Capitalist Societies', *Papers on Planning and Design*, Toronto: University of Toronto, Department of Urban and Regional Planning.

Sack, R. (1986) *Human Territoriality: Its Theory and History*, Cambridge: Cambridge University Press.

Sack, R. (1980) *Conceptions of Space in Social Thought*, Minneapolis: University of Minnesota Press.

Sanger, D. (1985) 'Star Wars Industry Rises', *New York Times* 19 November, Business Day, 25, 32.

Sartre, J-P. (1971 and 1972) *L'Idiot de la famille*, Paris: Gallimard.

Sartre, J-P. (1968) *Search for a Method*, translated by H. Barnes, New York: Vintage Books.

Sartre, J-P. (1960) *Critique de la raison dialectique*, Paris: Gallimard, translated by A. Sheridan-Smith (1982) *Critique of Dialectical Reason*, London: Verso.

Sartre, J-P. (1956) *Being and Nothingness: An Essay in Phenomenological Ontology*, translated and with an introduction by H. Barnes, Secaucus: Citadel.

Saunders, P., and Williams, P. (1986) 'The New Conservatism: Some Thoughts on Recent and Future Development in Urban Studies', *Environment and Planning D: Society and Space* 4, 393-99.

Saunders, P. (1981; second ed. 1986) *Social Theory and the Urban Question*, London: Hutchinson.

Sayer, A. (1985) 'The Difference that Space Makes', in Gregory and Urry (eds), 49-66.

Sayer, A. (1984) *Method in Social Science: A Realist Approach*, London: Hutchinson.

Sayer, A. (1982) 'Explanation in Economic Geography: Abstraction versus Generalization', *Progress in Human Geography* 6, 68-88.

Scheer, R. (1983) 'California Wedded to Military Economy but Bliss is Shaky', *Los Angeles Times* 10 July, part VI, 13, 14.

Scott, A. (1986) 'High Technology Industry and Territorial Development: the Rise of the Orange County Complex, 1955-1984', *Urban Geography* 7, 3-45.

Scott, A. (1984) 'Industrial Organization and the Logic of Intrametropolitan Location III: Case Studies of the Women's Dress Industry in the Greater Los Angeles Region', *Economic Geography* 60, 3-27.

Scott, A. (1983) 'Industrial Organization and the Logic of Intrametropolitan Location II: A Case Study of the Printed Circuit Industry in the Greater Los Angeles Region', *Economic Geography* 59, 343-67.

Scott, A. (1980) *The Urban Land Nexus and the State*, London: Pion.

Scott, A., and Storper, M. (eds) (1986) *Production, Work, Territory: The Geographical Anatomy of Industrial Capitalism*, Boston: Allen and Unwin.

Smith, M., and Feagin, J. (eds) (1987) *The Capitalist City*, Oxford: Basil Blackwell.
Smith, N. (1987) 'Dangers of the Empirical Turn: Some Comments on the CURS Initiative', *Antipode* 19, 59–68.
Smith, N. (1984) *Uneven Development*, Oxford: Basil Blackwell.
Smith, N. (1981) 'Degeneracy in Theory and Practice: Spatial Interactionism and Radical Eclecticism', *Progress in Human Geography* 5, 111–18.
Smith, N. (1980) 'Symptomatic Silence in Althusser: the Concept of Nature and the Unity of Science', *Science and Society* 44, 58–81.
Smith, N. (1979) 'Geography, Science and Post-positivist Modes of Explanation', *Progress in Human Geography* 3, 356–83.
Soja, E. (1988) 'Modern Geography, Western Marxism, and the Restructuring of Social Theory', in Peet and Thrift (eds), *The New Models in Geography*, Boston: Allen and Unwin.
Soja, E. (1987a) 'Economic Restructuring and the Internationalization of the Los Angeles Region', in Smith and Feagin (eds), 178–98.
Soja, E. (1987b) 'The Postmodernization of Human Geography: a Review Essay', *Annals of the Association of American Geographers* 77, 289–96.
Soja, E. (1986) 'Taking Los Angeles Apart: Some Fragments of a Critical Human Geography', *Environment and Planning D: Society and Space* 4, 255–72.
Soja, E. (1985a) 'The Spatiality of Social Life: Towards a Transformative Retheorization', in Gregory and Urry (eds), 90–127.
Soja, E. (1985b) 'Regions in Context: Spatiality, Periodicity, and the Historical Geography of the Regional Question', *Environment and Planning D: Society and Space* 3, 175–90.
Soja, E. (1984) 'A Materialist Interpretation of Spatiality', in Forbes and Rimmer (eds), 43–77.
Soja, E. (1983) 'Redoubling the Helix: Space-time and the Critical Social Theory of Anthony Giddens (Review Essay)', *Environment and Planning A* 15, 1267–72.
Soja, E. (1983) 'Territorial Idealism and the Political Economy of Regional Development', *City and Region: Journal of Spatial Studies* (Greece) 6, 55–73.
Soja, E. (1980) 'The Socio-spatial Dialectic', *Annals of the Association of American Geographers* 70, 207–25.
Soja, E. (1979) 'The Geography of Modernization: a Radical Reappraisal', in Obudho and Fraser Taylor (eds), *The Spatial Structure of Development*, Boulder: Westview, 28–45.
Soja, E. (1971) *The Political Organization of Space*, Washington DC: Association of American Geographers, Resource Papers.
Soja, E. (1968) *The Geography of Modernization in Kenya*, Syracuse: Syracuse University Press.
Soja, E., and Hadjimichalis, C. (1979) 'Between Geographical Materialism and Spatial Fetishism: Some Observations on the Development of Marxist Spatial Analysis', *Antipode* 11/3, 3–11.
Soja, E., Heskin, A., and Cenzatti, M. (1985) 'Los Angeles nel caleidoscopio della restrutturazione', *Urbanistica* 80, 55–60.
Soja, E., Morales, R., and Wolff, G. (1983) 'Urban Restructuring: an Analysis of Social and Spatial Change in Los Angeles', *Economic Geography* 59, 195–230.
Soja, E., and Scott, A. (1986) 'Los Angeles: Capital of the Late Twentieth Century', (Editorial Essay) *Environment and Behavior D: Society and Space* 4, 249–54.
Soja, E., and Tobin, R. (1974) 'The Geography of Modernization: Paths, Patterns, and Processes of Spatial Change in Developing Countries', in Brewer and Brunner (eds), *Political Development and Change: A Policy Approach*, New York: Free Press, 197–243.
Soja, E., and Weaver, C. (1976) 'Urbanization and Underdevelopment in East Africa', in Berry (ed.), *Urbanization and Counterurbanization*, Beverly Hills: Sage, 233–66.
Storper, M., and Christopherson, S. (1987) 'Flexible Specialization and Regional Industrial Agglomerations', *Annals of the Association of American Geographers* 77, 104–17.
Taylor, P. (1981) 'Geographical Scales in the World Systems Approach', *Review* 5, 3–11.
Thomas, W. (ed.) (1956) *Man's Role in Changing the Face of the Earth*, Chicago, University of Chicago Press.

Thompson, E.P. (1978) *The Poverty of Theory and Other Essays*, London: Merlin.
Urry, J. (1985) 'Social Relations, Space and Time', in Gregory and Urry (eds), 20–48.
Viehe, F. (1981) 'Black Gold Suburbs: the Influence of the Extractive Industry on the Suburbanization of Los Angeles, 1890–1930', *Journal of Urban History* 8, 3–26.
Walker, R. (1981) 'A Theory of Suburbanization: Capitalism and the Construction of Urban Space in the United States', in M. Dear and A. Scott (eds), *Urbanization and Urban Planning in Capitalist Societies*, 383–430.
Walker, R. (1978) 'Two Sources of Uneven Development under Advanced Capitalism: Spatial Differentiation and Capital Mobility', *The Review of Radical Political Economics* 10, 28–37.
Walker, R., and Storper, M. (1984) 'The Spatial Division of Labor: Labor and the Location of Industries', in Sawers and Tabb (eds), *Sunbelt/Snowbelt*, New York: Oxford University Press, 19–47.
Wallerstein, I. (1979) *The Capitalist World-Economy*, Cambridge: Cambridge University Press.
Wallerstein, I. (1976) 'The World System Perspective on the Social Sciences', *British Journal of Sociology* 27.
Weaver, C. (1984) *Regional Development and the Local Community: Planning, Politics and Social Context*, Chichester: John Wiley.
Williams, R. (1983) *Keywords: A Vocabulary of Culture and Society*, London: Fontana.
Wright, G., and Rabinow, P. (1982) 'Spatialization of Power: A Discussion of the Work of Michel Foucault', and 'Interview: Space, Knowledge and Power', *Skyline*, 14–20.
Zeleny, J. (1980) *The Logic of Marx*, Oxford: Oxford University Press.

# Index

259

# POSTMODERN GEOGRAPHIES
## The Reassertion of Space in Critical Social Theory
### EDWARD W. SOJA

Written by one of America's foremost geographers, this book contests the tendency, still dominant in most social science, to reduce human geography to a reflective mirror or, as Marx called it, an 'unnecessary complication'. Beginning with a powerful critique of historicism and its constraining effects on the geographical imagination, the author builds on the work of Foucault, Berger, Giddens, Berman, Jameson and, above all, Henri Lefebvre, to argue for a historical and *geographical* materialism, a radical rethinking of the dialectics of space, time and social being.

Soja charts the respatialization of social theory from the still unfolding encounter between Western Marxism and modern geography, through the current debates on the emergence of a postfordist regime of 'flexible accumulation'. The postmodern geography of Los Angeles, exposed in a provocative pair of essays, serves as a model in his account of the contemporary struggle for control over the social production of space.

**Edward W. Soja** teaches Urban and Regional Planning at the University of California, Los Angeles. He is the author of several books on African development and, more recently, on the economic and spatial restructuring of the Los Angeles region.

One of the most challenging and stimulating books ever written on the thorny issue of how and why societies use space for social purposes in the ways they do.
David Harvey, *Halford Mackinder Professor of Geography, University of Oxford*

ISBN 0 86091 936 6

**VERSO**
UK: 6 Meard Street London W1V 3HR
USA: 29 West 35th Street New York NY 10001-2291

Cover designed by Paul Wills
Illustration: *Good Morning City* by Hundertwasser (1969)